Cleaning Up the Environment

HAZARDOUS WASTE TECHNOLOGY

GREEN TECHNOLOGY

Cleaning Up the Environment

HAZARDOUS WASTE TECHNOLOGY

Anne Maczulak, Ph.D.

CLEANING UP THE ENVIRONMENT: Hazardous Waste Technology

Facts On File, Inc.
An imprint of Infobase Publishing
132 West 31st Street
New York NY 10001

Library of Congress Cataloging-in-Publication Data

Maczulak, Anne E. (Anne Elizabeth), 1954–
Cleaning up the environment : hazardous waste technology / Anne Maczulak.
p. cm. — (Green technology)
Includes bibliographical references and index.
ISBN 978-0-8160-7198-2
1. Hazardous waste site remediation. 2. Environmental responsibility. 3. Green technology. I. Title.
TD1052.M33 2010
628.5—dc22 2008042367

Facts On File books are available at special discounts when purchased in bulk quantities for businesses, associations, institutions, or sales promotions. Please call our Special Sales Department in New York at (212) 967-8800 or (800) 322-8755.

You can find Facts On File on the World Wide Web at http://www.factsonfile.com

Text design by James Scotto-Lavino
Illustrations by Bobbi McCutcheon
Photo research by Elizabeth H. Oakes

Printed in the United States of America

Bang FOF 10 9 8 7 6 5 4 3 2

This book is printed on acid-free paper.

Contents

Preface

The first Earth Day took place on April 22, 1970, and occurred mainly because a handful of farsighted people understood the damage being inflicted daily on the environment. They understood also that natural resources do not last forever. An increasing rate of environmental disasters, hazardous waste spills, and wholesale destruction of forests, clean water, and other resources convinced Earth Day's founders that saving the environment would require a determined effort from scientists and nonscientists alike. Environmental science thus traces its birth to the early 1970s.

Environmental scientists at first had a hard time convincing the world of oncoming calamity. Small daily changes to the environment are more difficult to see than single explosive events. As it happened the environment was being assaulted by both small damages and huge disasters. The public and its leaders could not ignore festering waste dumps, illnesses caused by pollution, or stretches of land no longer able to sustain life. Environmental laws began to take shape in the decade following the first Earth Day. With them, environmental science grew from a curiosity to a specialty taught in hundreds of universities.

The condition of the environment is constantly changing, but almost all scientists now agree it is not changing for the good. They agree on one other thing as well: Human activities are the major reason for the incredible harm dealt to the environment in the last 100 years. Some of these changes cannot be reversed. Environmental scientists therefore split their energies in addressing three aspects of ecology: cleaning up the damage already done to the Earth, changing current uses of natural resources, and developing new technologies to conserve Earth's remaining natural resources. These objectives are part of the green movement. When new technologies are invented to fulfill the objectives, they can collectively be called green technology. Green Technology is a multivolume set that explores new methods for repairing and restoring the environment. The set covers a broad range of subjects as indicated by the following titles of each book:

- *Cleaning Up the Environment*
- *Waste Treatment*
- *Biodiversity*
- *Conservation*
- *Pollution*
- *Sustainability*
- *Environmental Engineering*
- *Renewable Energy*

Each volume gives brief historical background on the subject and current technologies. New technologies in environmental science are the focus of the remainder of each volume. Some green technologies are more theoretical than real, and their use is far in the future. Other green technologies have moved into the mainstream of life in this country. Recycling, alternative energies, energy-efficient buildings, and biotechnology are examples of green technologies in use today.

This set of books does not ignore the importance of local efforts by ordinary citizens to preserve the environment. It explains also the role played by large international organizations in getting different countries and cultures to find common ground for using natural resources. *Green Technology* is therefore part science and part social study. As a biologist, I am encouraged by the innovative science that is directed toward rescuing the environment from further damage. One goal of this set is to explain the scientific opportunities available for students in environmental studies. I am also encouraged by the dedication of environmental organizations, but I recognize the challenges that must still be overcome to halt further destruction of the environment. Readers of this book will also identify many challenges of technology and within society for preserving Earth. Perhaps this book will give students inspiration to put their unique talents toward cleaning up the environment.

Acknowledgments

I would like to thank a group of people who made this book possible. Appreciation goes to Bobbi McCutcheon, who helped turn my unrefined and theoretical ideas into clear, straightforward illustrations. Thanks also go to Elizabeth Oakes for providing photographs that recount the past and the present of environmental technology. My thanks also go to Marilyn Makepeace, who provided support and balance to my writing life, and Jodie Rhodes, who helped me overcome more than one challenge. Finally, I thank Frank Darmstadt, executive editor, for his patience and encouragement throughout my early and late struggles to produce a worthy product. General thanks go to the publisher for giving me this opportunity.

Introduction

The industrial might of the United States grew in the 1930s and flourished during two world wars. As businesses large and small supplied the needs of the country, they discarded their wastes in landfills, pits, and waterways. It was an easy and inexpensive way to get rid of wastes before more were made. Waste dumping had not yet been linked with illnesses, so few people worried about it, especially if the materials were hidden under the ground or in the ocean. Within 20 years of the end of World War II, those wastes began to cause serious health problems.

From the late 1970s to the 1990s the government responded to the growing health threat by introducing new environmental laws. Each new law's main intent was to fix the hazardous waste problem. But laws are useless without the proper support. Part of that support would come from science. Fixing the waste problem required new technologies for testing the environment and for cleaning up *contamination.* This book explains how the world's air, land, and waters became polluted and how they are tested for pollution. Each chapter also describes the new methods being developed for removing dangerous chemicals from the environment.

Cleaning Up the Environment is about current methods and emerging methods in pollution cleanup. Hazardous wastes are removed from contaminated places by physical, chemical, or biological methods. These methods are described here, as well as their advantages and their disadvantages. The book also takes the reader through the entire step-by-step process of finding, testing, and cleaning up hazardous waste sites. It begins with contamination assessment and ends with a cleaned and restored body of land or water.

The first chapter follows the steps in contamination assessment. Assessment depends on accurate equipment for measuring chemicals in soil, air, or water. The methods and the very sensitive instruments now used for assessing contamination are described. The principles of good scientific data gathering are explained here.

Three chapters each describe in depth a method used today in pollution cleanup. The first is *excavation* using heavy equipment. Excavation may not be thought of as a cutting-edge method, but it plays a role in speeding up the removal of hazardous wastes that can threaten a nearby community. The second method is the biological cleanup of wastes. *Bioremediation* is pollution cleanup that uses microorganisms; *phytoremediation* is cleanup that uses plants. Both biological methods are described as well as the newest branches of each. The third method is chemical *oxidation* of hazardous compounds, which is one of the fastest growing remediation technologies today. Each method has found a place in remediating hazardous waste sites. This book also explains the advantages of a recent cleanup tactic: combining more than one of these technologies at a single hazardous waste site.

Two additional chapters study today's biggest cleanup tasks in the United States. One is the *brownfield* program, wherein hazardous waste sites are restored for human use or returned to the environment. The other is Superfund, which is this nation's program for cleaning up its very worst environmental messes. Each chapter devoted to these two subjects explores their history, the new technologies being used for cleanup, and their legal aspects.

Brownfields and Superfund sites almost always include water pollution. One chapter discusses water pollution and takes an in-depth look at civilization's reliance on a clean freshwater supply. Groundwater pollution receives extra attention for two reasons. First, Superfund sites usually cause groundwater pollution. Second, groundwaters supply much of the world with its drinking water, but their pollution has reached critical levels in many parts of the world. This chapter describes chemical and physical groundwater cleanup methods. It also covers international aspects of water pollution—especially the regions of the world where water demand is greater than water supplies can meet.

The goal of this book is to give the most up-to-date information on hazardous waste cleanup. Successful international programs are mentioned whenever they provide examples of problem-solving relevant to the United States. Staying abreast with new technologies is not always easy, but this book explains the basics of each technology. Readers can therefore use it as a reference for studying advances unfolding in environmental science.

Measuring Contamination

On the Earth's surface, the atmosphere, land, and water contain compounds that were not present in them centuries ago. This is because air, soil, and water *pollution* have reached into almost every part of the globe in the past 100 years. Today, sensitive equipment can detect tiny amounts of chemicals in the environment. Many scientists believe these instruments will soon show there is no spot on Earth free from pollution.

Before waste management experts can clean up large or small amounts of pollution, they must understand the specific substances that contaminate the environment. Contamination assessment has two purposes: to evaluate the overall pollution problem at a site and to measure the types and amounts of these hazardous substances.

Contamination assessment is done using equipment but also by simply observing a site. Sensitive equipment detects minute amounts of compounds, but the body's senses are also useful in evaluating contamination. People living within sight of factory smokestacks or alongside rivers covered by a toxic sheen need only their eyes and nose to detect pollution. Sight, taste, and smell often provide the first warnings of environmental dangers. Some contaminants, however, are not detectable by the senses, and analysis must be done on reliable equipment.

The antipollution laws set forth by the Clean Air and the Clean Water Acts of the 1970s could not be met without accurate equipment and trained scientists. Increase in knowledge about the chemicals in air, land, and water came by using improved machines that could measure various chemicals at low concentrations. The engineers who designed the equipment faced one challenge that is characteristic of environmental sampling: Polluting chemicals are usually part of complex mixtures. For example, a gram of

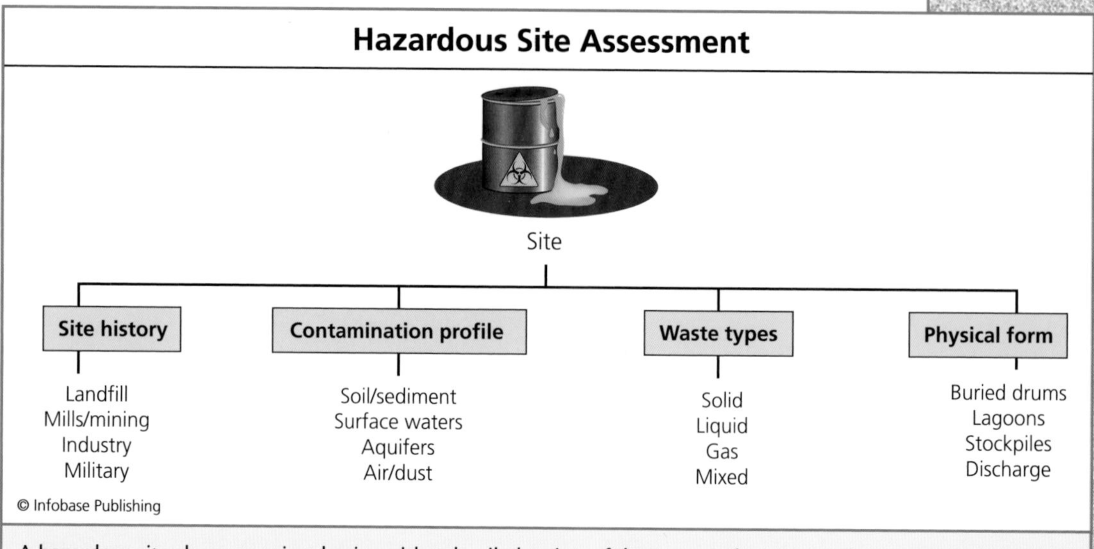

A hazardous site cleanup project begins with a detailed review of the types and amounts of contaminants that may be present on the site. Cleanup managers also research the history of the site to gain information on the chemicals that might have contaminated the area decades ago.

soil has thousands of compounds and many thousand living microorganisms. Environmental engineers had to design instruments capable of finding chemicals lurking at minuscule levels within these complex mixtures.

Scientists who study the environment by analyzing environmental samples view the world from the ecosystem level down to molecular and even atomic levels. They must track small amounts of chemicals because many pollutants cause harm to living tissue in amounts as low as one one-hundredth of a milligram (mg), so it is critical to have the ability to measure low concentrations. Contamination is now measured routinely to the molecular level in environmental laboratories. This chapter explores the planning and analysis steps in contamination assessment.

POLLUTION AND CONTAMINATION

The terms *pollution* and *contamination* are often thought of as having the same meaning. There is, however, a difference between them. Pollution is the presence of any solid, liquid, or gaseous form of energy harmful to biota, or living things, when it is in the environment. Pollution is chemical, biological, or physical. Chemical pollution is from compounds not naturally found in the environment. Biological pollution consists of microorganisms or nonnative plants or animal life. Physical pollutants are ionizing radiation, ultraviolet radiation, or heat (also known as thermal pollution). Some normally harmless forms of matter become pollutants only when they affect animal health. Pollen provides an example. Pollen is harmless at certain levels in the air but may irritate a person's eyes or throat at higher levels. When pollen becomes a health problem, it may then be defined as a pollutant. Contamination, by contrast, is matter in air, water, or soil that causes harm or death in an organism. Examples of contaminants are benzene in soil, viruses in water, and beta particles in the air. Because the definitions of pollution and contamination are so close, they will mean the same thing throughout this book.

The first step in waste management involves learning the characteristics of the pollution to be removed. Hazardous substances are dangerous in their own way and these characteristics are shown in the following table. But sometimes hazards change over time. For instance, many pesticides are slightly harmful to health, but as they decompose their breakdown products are more dangerous than the original compound. These breakdown structures are called intermediates. Thousands of intermediate compounds pollute the environment, and many of them are not well understood. Two things complicate the categories listed in the following table: First, intermediates may present different hazards than the original compound, but these hazards are unknown; second, pollution is seldom made of a single hazardous material. Most polluted sites contain mixed wastes. That is, they contain substances from two or more of these categories.

In addition to chemical hazards, waste management requires an understanding of the pollution's source and how it disperses from its source. Pollution comes from urban, agricultural, and industrial sources and there are subcategories within each of these. For example, the wastes released from mining, lumber operations, mills, shipping, and fishing are all forms of industrial pollution. Pollution disperses from its source in three main ways: through air, in water, or in soil. Of course, many pollutants move in all three. Air and water pollution create additional problems because they migrate to areas far from their source.

Another way to view chemicals polluting the environment is by whether or not they decompose in nature. Compounds that are decomposed in nature are *biodegradable.* Microorganisms found naturally in

TYPES OF HAZARDOUS SUBSTANCES		
TYPE OF SUBSTANCE	**PROPERTIES**	**EXAMPLES**
corrosive	capable of burning skin, eyes	strong acids, alkali metals
ignitable	capable of catching fire	propane
reactive	capable of exploding if exposed to certain compounds	chlorine gas, sodium, hydrogen sulfide
toxic	capable of causing harm inside the body	lead, mercury, pesticides, flame retardants
radioactive	capable of harming the body by emitting radioisotopes	uranium, nuclear fuel rods
etiological	capable of causing disease by a biological agent	bacteria, viruses, molds

soil and water carry out most of this decomposition. Nonbiodegradable pollutants cause more harm to ecosystems than biodegradable pollutants when they are ingested by wildlife and then enter food chains. The term *persistence* is another way of assessing a pollutant. Persistent compounds resist decomposition in nature and degrade over many years or not at all. They are also called recalcitrant compounds, or if they are *pesticides* they are called hard pesticides. *Recalcitrance* is related to persistence; it is the tendency of a compound to remain in the environment because microorganisms cannot easily degrade it. Some substances resist breakdown by microorganisms (also called *microbes*), but they do decompose, however, in the presence of light or at very high temperatures.

Regardless of how it is accomplished, fast decomposition in nature is better for ecosystems than pollution that persists for many months or years. This is because persistent compounds have a greater chance to be toxic to animal, plant, or microbial life. Many chemicals are not harmful in small doses for short periods of time, but are dangerous over a long-term, or chronic, period of exposure. Chemicals that stay in the environment also have greater opportunity to accumulate in food chains: this is known as *bioaccumulation*. Bioaccumulation occurs when plants or animals at the bottom of a food chain take up small amounts of a pollutant, and then these foodstuffs are eaten by another animal at the next step up the chain. As the pollutant becomes concentrated in animal tissues, animals at the higher levels get greater doses of the chemical. Bioaccumulation is discussed in more detail in *Pollution*, another volume of this set.

Some persistent items are not harmful; they simply ruin the enjoyment of a place. Though a mountain of old tires does not seem like pollution, it pollutes an environment by destroying the beauty of the nature around it. An environment free of pollution gives *intrinsic* benefits to people in the form of a clean and healthy place to live. Though it is not thought of as the environment, the indoors also exposes people to dangerous substances, as described in the "Harmful Indoor Chemicals" sidebar.

AIR, LAND, AND WATER

Pollution enters the air, soil, and water in a variety of ways. The common air pollutants are *volatile* compounds, gas emissions, ozone, small particles and fibers, heavy metals, acid rain, radioactive particles, and airborne infectious microbes. (Volatile compounds are those that escape into the air as a vapor.) Factory and incinerator smokestacks, exhaust pipes, construction sites, and dust-producing manufacturing sites are some of the many sources of air pollution. Natural events such as volcanoes can spew hazardous gases and dusts thousands of miles. Human-made disasters also contribute to the total amount of chemicals in the air. The massive destruction of New York City's World Trade Center on September 11, 2001, resulted in one of the nation's worst air pollution disasters. In the collapse, tons of lead, organic compounds, fine particles, smoke, and soot added to the amounts of these pollutants already present in the city's air. The disaster also released other materials not normally present in the air: jet fuel, cement, asbestos, foam, glass fibers, lint, charred wood and plastic, and paper fragments. Some compounds formed by the intense heat of the fireballs had never before been seen by chemists. "The stuff we saw in those plumes was truly nasty," said Thomas Cahill, professor of atmospheric science at the University of California–Davis. Taken together, chronic air pollution, natural events, and one-time disasters can make almost any substance airborne and extremely dangerous to animals if inhaled.

In polluted soil chemicals and particles migrate slowly compared with their movement in air or flowing waters. Microbes and plants therefore have a chance to detoxify many compounds. *Detoxification* is any set of reactions that make a hazardous substance less hazardous. But there is a disadvantage to this slow movement in soil. When the land holds a hazardous substance for long periods, the health of living things becomes threatened by chronic exposure to the substance. This is true, too, for bodies of water. Still, deep lakes are more threatened by chronic exposure to pollutants than are flowing waters. Today the sediments under lakes and

HARMFUL INDOOR CHEMICALS

Pollutant levels in indoor air are often several times greater than outdoors. Since the passage of the Clean Air Act in 1970, almost every U.S. city's air has become cleaner, but the same cannot be said for indoor air. Indoor air pollution comes from particles, smoke, gases, and chemicals. Nonliving particles may be dusts, fibers, asbestos, and crumbs from cooking, and biological particles consist of molds, pollen, and pet dander. In places where cooking is done by burning dung, wood, crops, or coal, indoor air pollution is a bigger health threat than outdoor air pollution. Building materials can emit *volatile organic compounds* (VOCs), many of which are associated with cancer in animals and humans and damage to the nervous system. Sources of VOCs are paints, varnishes, lacquers, waxes, cleaning solutions, disinfectants, cosmetics, and glues. Some house paints contain methylene chloride, which converts to unsafe levels of carbon monoxide. Additional indoor hazards are: radon from underground sources; nitrogen oxides from furnaces, stoves, and kerosene heaters; benzenes in smoke, paint supplies and vehicle fuels; formaldehyde in furniture, plywood, paneling, particleboard, high-gloss floors and cabinets, and foam insulations; styrene in carpets and plastics; chloroethylenes in dry-cleaned fabrics; lead in older lead-based paints and deteriorating buildings; and pesticides.

Energy-efficient buildings contain strong seals around windows and door frames, preventing good ventilation. Chemicals released from new carpeting, furniture, and plastics concentrate indoors, as do molds. The result is *sick building syndrome.* Its symptoms include dizziness, headaches, nausea, burning eyes, coughing, sneezing, fatigue, and irritability. Sick building syndrome has been difficult to explain in environmental medicine, but the following four prominent indoor at river bottoms hold a dangerous amount of heavy metals and pesticides that have settled there over many years. Fast-flowing rivers and the open sea have always been treated as places where hazardous substances disperse and become diluted. Environmental scientists now know that ecosystems in flowing waters are also damaged by pollution.

SAMPLING POLLUTED SITES

Accurate assessment depends on a representative sample, which is a sample that contains all of the constituents found in the environment and in the same proportions. Representative samples contain the following four features:

air pollutants probably contribute to it: cigarette smoke, formaldehyde, radon gas, and very small airborne particles, such as polyester fibers. Other harmful pollutants include molds, emissions from household products and pesticides, and carbon monoxide. Recent work by environmental health scientist Michael Apte of the Lawrence Berkeley Livermore Laboratory in California suggests that high outdoor ozone levels also worsen indoor air pollution. In 2008 *Scientific American* magazine quoted Apte as explaining, "We found that outdoor air pollution, ozone, is associated with symptoms of lower-respiratory and upper-respiratory stress that occur in buildings to workers." In other words, subjects in Apte's studies experienced worse symptoms indoors when ozone levels were high outdoors. Ozone may actually react with certain air filters meant to clean the indoor air and make conditions worse. Says Apte, "There is a six times greater likelihood that these symptoms will occur if you have both higher ozone levels and the polyester or synthetic filters."

Homeowners can reduce indoor air pollution by using the following items: vents from the kitchen, bathroom, and laundry room to the outdoors to prevent moisture buildup and mold growth; carbon monoxide monitors on gas and oil furnaces; new building materials, which emit low levels of *off-gases;* and heating systems with air-to-air heat exchangers that rely on fresh air from outdoors. The simple opening of windows and doors for ventilation and the use of vents on air conditioners and fans also help reduce the buildup of indoor air pollutants.

The U.S. Environmental Protection Agency (EPA) sets outdoor limits for carbon monoxide, nitrogen dioxide, sulfur dioxide, ozone, particles, and lead, but indoor air has no such limits. The indoors may be thought of as the new frontier of air pollution prevention.

- They come from the site that is to be cleaned up.
- They are large enough to allow completion of all testing.
- They contain replicates.
- They are reproducible.

(Reproducible samples are those from which other scientists get the same results if they analyze them by the same procedures.)

An analyst should describe each environmental sample taken from a particular place in the environment. For example, a sample from a lake might be described as one in which five fluid ounces (148 mL) were collected one foot (0.3 m) below the surface and 150 feet (46 m) from the eastern shore. Samples should also be drawn at random and from the entire test site. Random samples are samples that do not follow any pattern. (A person who avoids getting dirty feet in a muddy field and samples at only the dry spots is not taking a random sample.) Consider an acre of land from which an analyst is told to take 30 soil samples. In this procedure, $n = 30$, meaning 30 separate samples will represent the entire plot of land. The analyst takes random samples from every one of 30 sections of the entire acre. If all 30 samples were taken from only one section, the sampling is biased and it will not give a true picture of the pollution. Biased sampling is not part of good science because it leads to results that are not representative of the entire test site. Analysts follow the rules of sound science whenever they take samples from polluted sites and analyze in a laboratory. These guidelines are described in the "Principles of Good Science" sidebar on page 12.

Soil sampling tools range from shovels and small handheld probes, which reach no deeper than a few feet (about 1 m), to large machine-driven probes that collect soils as deep as 175 feet (53 m). Cylinder-style probes remove a column of earth called a core sample. Core sampling is also used for studying glaciers and the polar caps.

The pore water method works best for sampling deep sediments. Clean water is forced through tubes into the spaces between sediment pieces. The water then accumulates in collection wells and passes through a thin, porous filter, also called a membrane. In the membrane diffusion method, small organic compounds diffuse out of sediments and cross the membrane. Environmental scientists also use sensors containing diffusion membranes for measuring the amounts of soil contaminants.

(a) Manual samplers take core samples of soil from a few feet below the surface. *(Geotechnics Sales Division)*

(b) Large sampling equipment reaches soils deep below the ground's surface and sediments near groundwaters. *(Soil Essentials)*

Scientists take representative water samples as close to pollution sources as possible and a few feet below the surface. Sample sizes range from two fluid ounces (59 mL) to more than 200 gallons (750 l). Small samples are collected by hand in clean, nonbreakable bottles, while larger samples require hoses and pumps. Because water is a flowing system, many replicate samples should be taken over time to account for changes in conditions.

For measuring air pollution, portable air-sampling machines draw in a set volume of air. Some samplers analyze the chemicals on the spot. Others collect the air on an *adsorbent* material and the sample is then analyzed in a laboratory.

Professionals who visit polluted sites to assess the site's conditions prefer portable sampling devices that also analyze the samples on the spot. Pollution detection equipment now detects various substances in water, wastewater, hazardous wastes, and air. Some of the commonly used portable devices are listed in appendix A, with a brief description of their features.

ANALYTICAL TOOLS AND INSTRUMENTS

Environmental measurements are of three main types: gravimetric, biological, and physicochemical. Gravimetric methods are those that measure the mass of a substance. The weight of a 0.00001-gram particle of soot as measured on a laboratory balance is a gravimetric result. Biological methods are used for measuring living species in a sample. Biological contamination is expressed as the number of units (cells or organisms) per volume or gram of sample. Physicochemical methods measure units other than mass or biological units. They measure either physical characteristics or chemical properties of a hazardous substance.

A physicochemical instrument may be a simple pH meter that measures acidity, or a more intricate piece of equipment. Physicochemical instruments belong to groups according to the type of analysis they perform: chromatography, colorimetry, fluorometry, nuclear radiation counting, polarography (voltammetry), or spectroscopy (spectrometry). Some analysis methods are described in the table on page 14.

Chromatography is a means of separating a compound from other compounds and was one of the first techniques used by the pioneers of analytical chemistry. It would be useful, for instance, in separating a pesticide from many other substances collected in a sample from a river. For this reason, chromatography is often combined with other methods that identify the chemical once it has been separated from a group of chemicals. Each of the methods described here are being improved upon almost daily to measure ever tinier amounts of contaminants in the environment. Most analyses measure a contaminant's concentration, which is the amount of contaminant within a known amount of another substance. For example,

PRINCIPLES OF GOOD SCIENCE

All well-planned scientific experiments address a null hypothesis, which is a theory believed to be true. In environmental chemistry, an example of a null hypothesis is the following: Brownfield site ABC is contaminated with hazardous levels of the pesticide, heptachlor. First, a chemist proposes a null hypothesis, then uses the principles of good science to prove or disprove it.

Scientific measurements must be accurate and precise in order for an analyst to trust them. Accuracy is the correctness of a scientific result—how near it is to the true value. For instance, an air analysis that shows the atmosphere to be 78 percent nitrogen and 21 percent oxygen is accurate, because many previous studies have determined this composition of the air. Precision is the exactness of a scientific analysis—how well a set of replicate measurements agree with one another. A technique that measures mercury in water in amounts as small as parts per billion is an example of precision.

Scientific findings are of no use to other scientists unless they have been written down. Documentation is one of the five basic components of good science, which are (1) proper training; (2) written procedures before testing begins; (3) data collection; (4) documentation; and (5) reporting the results. Before collecting samples, it is important also to have a clear idea of what will be measured and how it will be measured.

The analyst's data are always checked for trustworthiness by two methods. These are quality assurance and quality control, or QA/QC. Quality assurance is the review of activities to make certain that the analyst follows the written procedures. Quality control involves checking the data and calculations for accuracy.

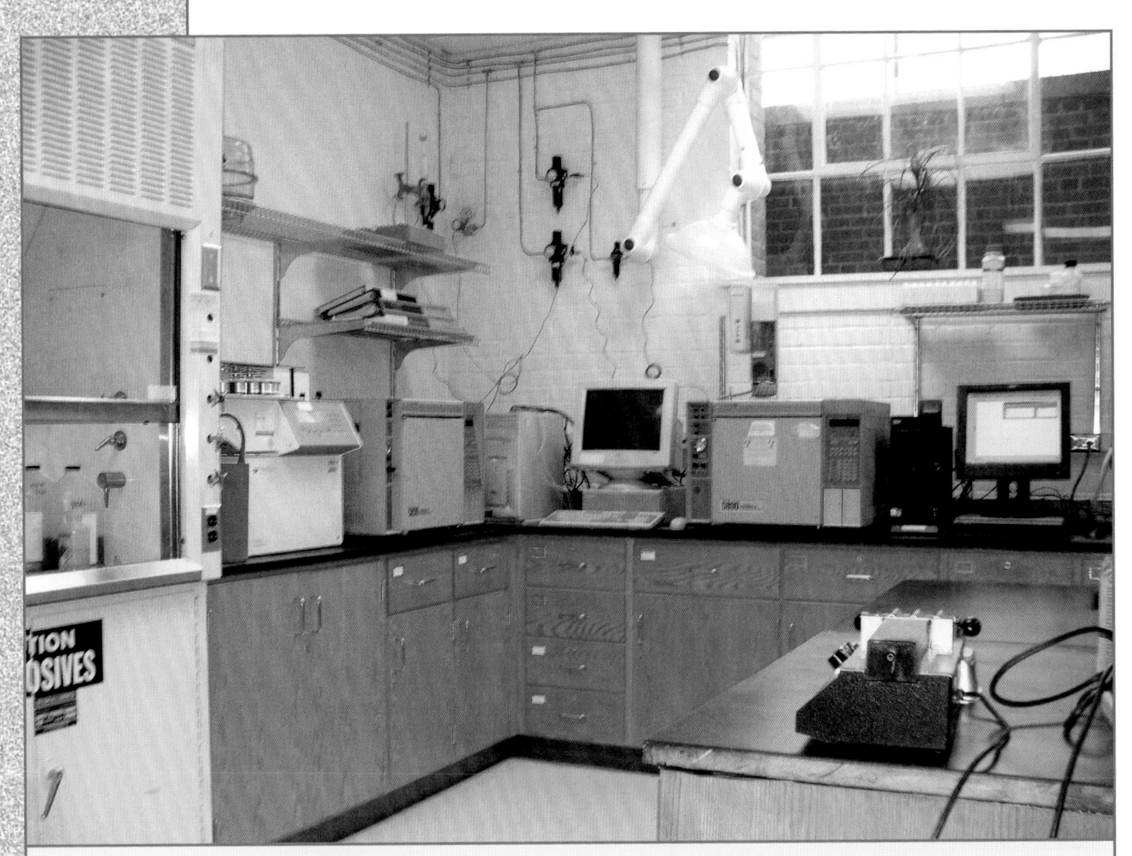

Analytical laboratories use a variety of chemical, physical, and biological methods to detect the presence of pollutants in air, soil, and water. New instruments have enabled chemists to detect pollutants in parts per trillion levels, and analysis will soon reach to the parts per quadrillion level. *(Robert C. Borden, North Carolina State University)*

mg/L, or milligrams per liter, is a unit of concentration of a liquid, and *ng/square foot,* or nanograms per square foot, is a unit of concentration of an air sample.

The development of today's sophisticated analytical methods began between 1848 and 1914. During this period a handful of industrialists gained an understanding of the potential benefits and profits that chemistry offered. In Germany chemist Adolf von Baeyer designed new compounds he hoped would improve human health, and by doing so he introduced the new field of synthesis chemistry. Other scientists such as Englishman William H. Perkin used their fascination with chemical structures to help a burgeoning coal industry. This industry needed ways

TYPES OF PHYSICOCHEMICAL ANALYSIS	
TYPE OF ANALYSIS	**WHAT IT MEASURES**
colorimetry	unique color emitted by a substance and converted into a number
spectrophotometry (visible, fluorescent, ultraviolet, or infrared light)	same as colorimetry but more sensitive and uses more of the light spectrum
atomic absorption spectrophotometry	light absorbed when a chemical is ionized
turbidimetry	light-blocking of a solution to determine concentration of particles
spectroscopy (also called spectrometry)	relationship between compound and light or other nonnuclear radiation
fluorometry	fluorescence emitted by a substance when exposed to ultraviolet light
chemiluminescence	light emitted from a chemical reaction
polarography (a type of voltammetry)	amount of electric current conducted by a compound
nuclear radiation counting	amount of nuclear particles emitted from radioactive matter

to make use of tons of by-products from its operations, especially coal tar. Coal companies recruited Perkin, his son, W. H. Perkin, Jr., and others to find ways for extracting useful compounds from coal tar. These scientists depended on increasingly sensitive and fast ways to analyze the new compounds they purified in their laboratories.

At the turn of the century German chemist Karl Fresenius set up an analytical laboratory devoted to identifying samples sent to him from police departments and chemical companies, as well as coal companies. There seemed to be no shortage in work as new industries based on chemistry grew: industrial acids and bases, soaps, fermentation products, and drugs. Even the military sent new types of explosives to Fresenius for testing in advance of the coming Great War (World War I). Then as now, industries wanted to make products of a standard quality for their customers. They could do this only by analyzing their products and learning the exact composition of every batch they produced.

In the early to mid-1900s scientists refined the techniques for studying elements and atoms rather than whole compounds. By doing this they gave birth to a new field called quantum chemistry. Max Planck made quantum chemistry's first major discovery when he related wavelength to energy, yet his professors tossed aside his doctoral dissertation, partly because Planck's theories had never before been proposed. Planck struggled for years to win acceptance from his peers until a technician employed in a Swiss patent office took notice. His interest grew regarding Planck's idea of invisible "energy packets." He set up a series of experiments on these packets, or atoms, and soon built on Planck's concept to define light as a wave of particles. Physicists the world over found this scientist, Albert Einstein, and his ideas odd at best. But the idea of the atom was to become accepted as more physicists and a mathematician by hobby, Niels Bohr, continued their attempts to describe the structure of atoms. Bohr described quantum numbers as the electron energy levels, called orbits, surrounding each nucleus. Much of today's analytical science is now based on Bohr's model of electron movement from one energy level to the next.

Spectroscopy is founded on the relationship between matter and the energy in light. In environmental science, the matter being measured is contamination, and each contaminant releases its energy as light. The light is then converted to a measurable color. Three types of spectroscopy carry out this conversion: scattering, emission, and absorption. Scattering spectroscopy measures the amount of light redirected by a substance. Emission spectroscopy measures light energy released from molecules (or atoms) when dropping from a high energy level to a lower energy level. Absorption methods detect a molecule's energy change as it absorbs electromagnetic energy, which is electricity within a magnetic field. New specialized equipment now measures the scattering, emission, or absorption of ultraviolet light, infrared light, X-rays, gamma rays, and laser light, and spectroscopists today examine molecules, atoms, and electrons.

As mentioned earlier, chromatography separates compounds from one another. The most powerful analyses combine chromatography with

PHYSICOCHEMICAL INSTRUMENTS

DEVICE	PRINCIPLE	ANALYSIS
GC	vaporized compounds divide between mobile gas and immobile liquid	organic compounds
GCMS	separates mixtures of compounds and then identifies compounds	organic compounds, hydrocarbons, VOCs
HPLC	separates, identifies, quantifies soluble compounds	aromatic hydrocarbons, carbohydrates
O_2 demand	oxygen uptake by microbes or chemicals	proteins, carbohydrates, fats
AA	measures energy levels of vaporized atoms or ions	metals
IPC/MS	ions from compounds are identified	elements

Note: GC = gas chromatograph; GCMS = gas chromatograph-mass spectrometer; HPLC = high-performance liquid chromatograph; AA = atomic absorption spectrometer; IPC/MS = inductively coupled plasma mass spectrometer.

sensitive detectors to find specific compounds after they have been separated from others. For example, mass spectrometers detect low levels of chemicals. The mass spectrometer attaches to a gas chromatograph (GCMS) or a high-performance liquid chromatograph (HPLC-MS) for very precise measurements.

The laboratory instruments used in pollution analysis usually are named after the method of analysis they perform. For instance, a GC, or gas chromatograph, performs gas chromatography, a method that measures compounds after they have been vaporized. The main analytical devices are listed in the table above according to the methods they perform.

Analytical methods should possess both high resolution and high sensitivity. Resolution is the ability of an instrument to distinguish between different compounds, and sensitivity is the minimum level the instrument can detect. Equipment has improved greatly in these two areas since the passage of the Clean Air and Clean Water Acts. Compounds measured to mg amounts in the 1980s are now detected in nanogram (ng) amounts; a nanogram is a billionth of a gram. In addition, analytical methods find certain types of compounds, even tiny amounts vaporized in air. Examples of the various instruments available for analyzing air are listed in the following table.

Physicochemical instruments measure exact amounts of a chemical. Biological methods use living systems to test the harmful effects of a chemical in living things. Results from biological tests tend to be expressed in a range of values rather than an exact value produced by an instrument. Biological testing therefore seems to be less sensitive than laboratory

Air Pollution Measurement Technology

Pollutant	Common Analysis Method	Level of Sensitivity
nitrogen oxides	chemiluminescence	0.01 mg/L
sulfur dioxide	fluorescent spectrophotometry	0.01 mg/L
carbon monoxide	infrared spectrometry	1.0 mg/L
ozone	ultraviolet spectroscopy	0.01 mg/L
heavy metals	atomic absorption spectrometry	0.01 mg/L
particles	particle counter	0.1 μm diameter
volatile organics (VOCs)	chromatography-spectrometry	1.0 μg/L

Note: mg = milligram; μg = microgram; μm = micrometer

BIOLOGICAL TESTING OF POLLUTANTS		
TEST ANIMAL	ANIMALS IT REPRESENTS	ECOSYSTEM
minnow	fish	freshwater
water flea	crustaceans	freshwater
silverside fish	fish	estuary
marine shrimp	fish and crustaceans	marine (salt water)

instruments. But biological tests are valuable because they relate pollution to ecosystems. In biological testing, a scientist puts a known amount of a chemical contaminant into a tank containing minnows, insects, or mollusks. The scientist then monitors the animals for any signs of harm to their health or behavior. The table above gives examples of the most widely used biological test systems.

Biologists develop new methods far more sensitive than today's most powerful instruments. One such method uses biological sensors, or *biosensors.* A biosensor combines two microbial enzymes on a single probe. One enzyme called the receptor activates in the presence of a pollutant such as *DDT.* The second enzyme, the reporter, converts the reaction into a visible signal. The luciferase enzyme found in a number of bacteria—it is the compound that makes fireflies light up at night—has been used in biosensors, which light up in the presence of a specific hazardous substance. A technician simply immerses the biosensor's probe in a soil or water sample and it lights up within minutes if DDT is present.

Polymerase chain reaction (PCR) probes detect nucleic acids, such as deoxyribonucleic acid (DNA) and ribonucleic acid (RNA). For example wastewater overflowing onto beaches pollutes the coastal waters with disease-causing viruses and bacteria called *pathogens.* PCR detects very small quantities of these pathogens on the beach or in the water and, like biosensors, they work quickly.

Each biological or electronic method has a level of detection (LOD). LOD is the lowest level that a method or an instrument can measure. Many of the contaminants monitored in the United States and the United Kingdom must be at levels of mg/L or lower to be considered safe to biota. These low concentration levels are explained in the "Parts

Parts per Million and Parts per Billion

Many pollutants are dangerous in minute amounts in the living tissue of humans or animals. Environmental scientists therefore measure hazardous chemicals in units called parts per million (*ppm*), parts per billion (*ppb*), and even parts per trillion (*ppt*). These are equal to mg/L, µg/L, and ng/L, respectively.

Polluted water sometimes contains large amounts of sediments or salts. Either of these changes the mass of a gram of water, so *ppm* must be converted to a value that accounts for the change in water's density:

$$ppm = \text{mg/L} \div \text{specific gravity of fluid}$$

Real-life examples are useful in illustrating these small concentrations. One *ppm* is equivalent to one drop of water in a 15-gallon (57-l) trash can filled with water. One *ppb* is one kernel of corn in a 45-foot (13.7-m) tall, nine-foot (2.7-m) diameter silo filled with corn. One *ppt* is one square foot (0.09 square m) of land in the entire state of Indiana.

per Million and Parts per Billion" sidebar above. As an example, the U.S. limit for arsenic in drinking water is 0.01 mg/L, equal to 0.01 parts per million.

TYPES OF CONTAMINATION CLEANUP

Contamination cleanup begins with assessment using eyes, nose, physicochemical instruments, and biological probes. Once an investigator determines the extent of the pollution by any or all of these approaches, a cleanup plan must be developed for the hazardous site. The cleanup team has many technologies at its disposal, but picking one is not always a straightforward job because hazardous substances can take many forms. A warehouse full of chemical-filled storage drums is a hazard, but not in the same way as an outdoor waste pit. Likewise the cleanup technologies for a stockpile of industrial chemicals dumped behind a factory differ from those needed to restore a town's contaminated drinking water. In

(continues on page 22)

CASE STUDY: BIRTH OF THE U.S. ENVIRONMENTAL PROTECTION AGENCY

The seeds of the United States Environmental Protection Agency (EPA) date to the mid-1800s during a time in which the country's industries grew, generating greater amounts of wastes as production swelled. Water pollution threatened the public's health as early as the 1880s, both in industrialized cities and agricultural areas. By the 1920s large cities such as New York, Boston, and Philadelphia began forming public health boards to devise plans for controlling the illnesses caused by tainted water and poor air quality. No U.S. laws existed at the time, however, to protect the environment or to hold polluters responsible for any illness their wastes caused. Factories and feedlots discharged and dumped wastes wherever they could, and with no penalty. Industry and government leaders yearned for the United States to become a manufacturing giant, so wastes had become viewed as a regrettable but necessary by-product of this ambition. Unbeknownst to most people, the environment was approaching a limit in its capacity to hold increasing amounts of waste.

An incident illustrating the looming catastrophe occurred in 1948 in the mill town Donora, Pennsylvania, near Pittsburgh. For three consecutive days emissions laden with sulfur dioxide from a local metals plant filled Donora's sky. Mayor John Lignelli described the conditions in Donora as the smog cloud rolled over the town while he and others sat at a local football game. "You couldn't identify the ballplayers," he recounted for the *Pittsburgh Post-Gazette* in 1998. "You could see movement on the field, but you didn't know who had the ball and what was going on. But we stayed and watched." Six thousand people in Donora soon developed pulmonary disease; twenty died.

After the air cleared, no person or company took responsibility for the health hazard. A handful of forward-thinking citizens had already been worried about the decaying air quality for years, but they felt their voices went unheard even after the disaster in Pennsylvania. In fact, the Donora environmental disaster began the building of momentum toward control of air pollution. In 1963 in New York City, thousands of people became sick and 300 died from illnesses caused by pollution in the air above their rooftops. Environmentally conscious citizens finally felt enough power to pressure Congress into passing the Air Pollution Act. But environmental damage had already grown to crisis levels. In 1969 in Cleveland the oil-polluted Cuyahoga River caught fire and the oil burned for eight days. A short drive away, fish died by the millions and Lake Erie's beaches closed, the water too toxic for swimming. At the 50th anniversary of the Donora event, the EPA's associate director Marcia Spink acknowledged "the debt

of gratitude the people of the United States owe Donora and the event that led to the federal Clean Air Act of 1970."

The country's young environmental movement put increasing pressure on leaders in Washington, D.C., to stop ignoring the pollution problem. In 1970 President Richard Nixon hastily established the EPA, perhaps more as a political decision than an environmental one. Upon signing the act, however, the President seemed to have understood the voters' wish for a healthier environment. Nixon stated that "the 1970s absolutely must be the years when America pays its debt to the past by reclaiming the purity of its air, its waters, and our living environment. It is literally now or never." The agency's newly appointed administrator, William D. Ruckelshaus, took on a massive task; employees of the new agency possessed little training in environmental science, yet it accepted its new responsibilities by targeting the following two objectives:

1. oversee programs for removing pollutants from air, soil, water, and buildings
2. provide pollution-prevention guidance to citizens, industry, and local governments

The new agency began operating with almost no blueprint as to how these goals were going to be achieved, yet Ruckelshaus gave his staff a confident proclamation in a December 4, 1970, memorandum. He urged them to proceed "with the valuable work, which is already underway. We cannot afford even a slight pause in the ongoing efforts to preserve and improve our environment."

Throughout the 1970s the administrator bolstered the EPA's ranks with trained scientists and specialists in the environment and ecosystems. As the agency grew to its current 18,000 employees, experts in government, law, project management, health, and education joined the scientists. Today's EPA is headquartered in Washington, D.C., and has 10 regional offices serving all the states and U.S. territories, and coordinates all activities related to improving the environment. To carry out its goals, the EPA sets and enforces pollution standards, and it fines companies or individuals who do not follow environmental laws. It offers guidance to the public on how to control pollutants. For instance, the EPA Web site provides the current listing of hazardous chemicals and the highest amounts of them allowed by law to be present in the environment, and also

(continues)

(continued)

lists the status of all the Superfund sites in the United States and its territories. The EPA also oversees the endangered species list, participates in programs to halt global warming, and it supervises the nation's drinking water quality, industrial chemicals, pesticides, air quality, and radiation. Finally, this agency cooperates with other federal agencies on health issues and homeland security.

The U.S. Environmental Protection Agency has since 1970 had the responsibility to safeguard the public from environmental hazards. Part of the agency's duty involves enforcing the latest U.S. laws protecting air, land, and water quality. *(EPA)*

Such monumental tasks for protecting the environment and carrying out all its other jobs come with plenty of obstacles. The president of the United States appoints the EPA's director, with confirmation by the senate, and many critics believe this arrangement leads to a difficult balancing act between industry goals, politics, and environmental concerns. Many in the public often wonder how well the EPA is looking out for them, yet some industries voice frustration over EPA mandates. The EPA truly has one of government's most difficult challenges.

(continued from page 19)

other words there are many types of cleanup methods for the many types of contamination that might be found at a hazardous waste site.

Current cleanup technologies aim to minimize the spread of contaminants during the cleanup process. It is an advantage to treat contaminants at the cleanup site as soon as they have been excavated, extracted, or pumped from the ground. This is because on-site cleanup combined with treatment reduces the need to haul hazardous materials to other locations.

Costs can be lowered, and immediate on-site treatment also lowers the health risks to people and animals living close to hazardous wastes.

Today's contamination assessment can be summarized as having the following two main components:

1. analysis of the contamination
2. selection of the best cleanup/treatment method

The EPA provides guidance for all of these considerations when a company must clean up a site. "Case Study: Birth of the U.S. Environmental Protection Agency" on page 20 describes the responsibilities of this large government agency.

INTERNATIONAL PROGRAMS IN WASTE CONTROL

Every country has a stake in preventing pollution for a simple reason: Waste does not stay in one place. Global shipping and travel spreads hazardous matter farther and faster than ever before, and pollutants in the oceans and in the atmosphere affect the health of people in distant places. Contaminants from afar limit the amount of land and water available for people's use in every country. As these resources become scarcer, the affected country has less land for food production and less dependable drinking water. The risk of conflict then increases, especially if the pollution is coming from another country. Conflicts and war over natural resources have occurred for centuries. Today people fight over oil, minerals, precious gems, gold, fishing grounds, and other natural resources. Competition for water and land will increase as urban populations expand outward. Unlike oil or salt or spices—all causes of war at least once in history—water and land will never lose value because the human species needs them to survive.

Many international organizations try to help countries avoid conflicts over natural resources. They work with governments to design methods for controlling pollution as well as devise ways to conserve natural resources. In the process nations agree to cooperate and share scientific knowledge about the environment, natural resources, and health for the benefit of everyone.

Pollution rarely stays in one place. International agreements help countries cooperate on limiting the amount of pollution they put into the environment. The United Nations plays a role in monitoring the status of global pollution. *(iStockphoto)*

One example of an international organization that sponsors environmental studies is the United Nations Environment Programme (UNEP). UNEP advises nations in areas of toxic chemical management, marine and coastal waters, oil pollution, greenhouse gases, and other environmental concerns. Nations abide by UNEP bans on ocean dumping and *incineration* at sea. In 1976 UNEP established the International Register of Potentially Toxic Chemicals, an information database of over 8,000 hazardous chemicals used worldwide. Member nations refer to this database and other UNEP programs when they assess their own contamination issues.

Some international organizations focus on specific topics such as oil spill cleanup or hazardous-waste transport. Whether providing general or specific advice, the organizations listed in appendix B supply storehouses of information on the Earth's environment.

CONCLUSION

Contamination assessment is the determination of the amounts and types of pollutants in a specific place. Before pollution cleanup begins, the site's air, soil, and water must be analyzed for the presence of hazardous substances. Without contamination assessment, it would be impossible to design a cleanup plan. Assessment begins with an evaluation of the form of the pollution; pollution may be in air, soil, or water, and in the form of gases, solids, or liquids. Following this overall assessment, the types of compounds and their amounts must be measured. This requires specialized equipment to separate one chemical from many others before analysts can identify it. All of the procedures done at the site or in a laboratory must follow sound scientific principles. Accurate measurements rely on representative samples, meaning that each sample contains the same pollutants at the same amounts as found in the location where it was taken. Accuracy, precision, and proper recordkeeping are essential for producing reliable data.

New methods in chemical analysis developed during the industrial growth of the 19th century. Many of these methods have been adapted for measuring environmental samples. New technologies now detect small amounts—ppm to ppt—of pollutants in places previously thought to be clean. In the United States, polluters must follow environmental laws and the EPA provides cleanup teams with resources on measuring contamination and treating it. International organizations conduct programs similar to the EPA's, but they focus on cooperation between nations in solving pollution problems. Several international agencies provide guidance to their member nations on general or specific topics in environmental protection.

Instruments improve almost daily to detect ever smaller amounts of contamination. All of these measurements depend on accurate and precise analyses and the principles of good science.

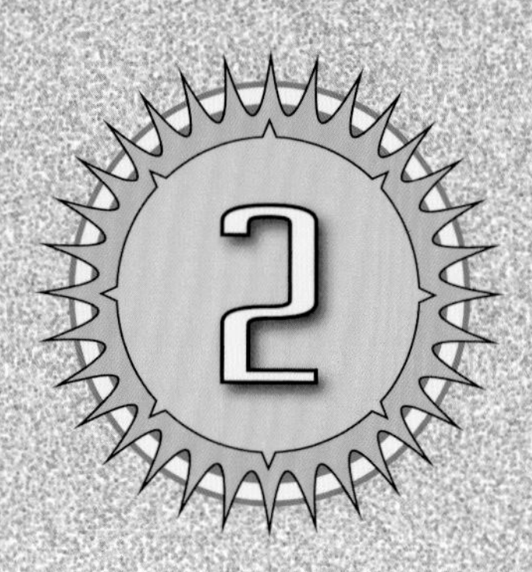

Excavation of Contaminated Sites

It seems there can be no simpler approach to cleaning up pollutant-defiled land than to pick up the polluted soil and haul it away. This is the cleanup process called *excavation.* A fleet of backhoes, bulldozers, and trucks run by trained operators serve as the fastest and cheapest way to excavate and get rid of pollution. Excavation projects of many years ago followed rather sloppy techniques compared with those done today. Large equipment simply rumbled onto sites, scooped up tons of soil, and dumped it into trucks for transport to a disposal site. Cleanup crews probably left behind a good portion of the pollutants.Today's excavation involves more careful procedures, but it remains a coarse way to remove hazardous wastes. For this reason hazardous waste experts often refer to excavation as "dig-and-dump."

As environmental scientists learned more about hazardous substances, they realized pollution often percolated into soils deeper than the topsoil and reached groundwaters, which are natural underground water sources. Environmental studies also showed that different contaminants behave in different ways in the environment. Scientists learned the value of monitoring contaminants before, during, and after a cleanup. If excavation were to remain a reasonable approach to contamination cleanup, something better than dig-and-dump would be required.

Contamination assessment takes place before any excavation begins. The cleanup project's manager determines the type of hazardous substance (or mixture of substances) present before setting up a plan for excavating it. The U.S. Environmental Protection Agency (EPA) helps by divid-

Superfund sites are places designated by the U.S. Environmental Protection Agency as containing the highest levels of hazardous chemicals. Many Superfund cleanups require the cleanup of both soil and water at the site. This Superfund site in Macomb County, Michigan, was a hazardous liquid disposal site. *(Michigan Department of Environmental Quality)*

ing hazardous substances into categories based on their chemical form: mercury wastes, munitions wastes, radioactive materials, phenols, and chlorinated compounds. The EPA further provides guidelines on how to clean up sites polluted with specific chemicals. It offers online cleanup tips for the following chemicals: dioxins; pesticides; asbestos; lead; mercury; polychlorinated biphenyls (PCBs); volatile organic compounds (VOCs); methyl tertiary-butyl ether (MTBE); and paint wastes. This information helps cleanup teams create the best excavation plan.

Today's excavation projects include safeguards to prevent the spills and losses that used to take place during excavation. For instance, the EPA requires the cleanup team to monitor contaminant levels throughout the process. In addition the team must consider the dangers specific to the polluted site. Hazardous waste sites near flood zones or on earthquake fault lines require special planning, for example. Excavation crews now follow strict safety procedures to protect them from being harmed by contaminants. At the same time, they follow procedures for preventing spills and keeping contaminants from dispersing into the surroundings.

Excavation serves well when more advanced cleanup methods are too expensive or not appropriate for a contaminated site. Excavation was once the cheapest cleanup method used in the United States, and though the cost of new sophisticated techniques is decreasing, excavation remains the fastest, cheapest way to clean up a large quantity of contaminants. Speed is especially critical when the hazardous wastes occur near schools, offices, or homes.

This chapter reviews excavation procedures, the rocky history of some excavation sites, and emerging technologies expected to replace dig-and-dump.

THE EXCAVATION PROCESS

Site cleanup always starts with observation. Discolored dirt, lack of vegetation, oily puddles, or piles of ash and debris all provide clues of where an excavation job should begin. From the rise of industrialization to the early 1980s many companies discarded waste chemicals directly on the ground outside their manufacturing plants. Congress began strengthening environmental laws in the 1980s and one by one individual states started to enforce stricter laws for handling and disposing of industrial chemicals. Quite a few companies that had been dumping their wastes on their own property could not meet the requirements and went out of business. Military bases had also left behind hazardous wastes and discarded munitions. Both industrial and military sites therefore became the bulk of today's brownfield and Superfund projects, discussed in chapters 5 and 7, respectively. Excavation seemed the most straightforward way to remove the tons of hazardous materials left behind by defunct companies and decommissioned military bases.

Today's excavation jobs begin with an inspection of the site to determine the size of the waste load. Aboveground piles of hazardous waste are easy to see, but hazardous materials (hazmat) technicians also take samples to find compounds hidden in the soil so that a cleanup manager has a clear picture of the entire problem. Hazmat technicians determine the distance contaminants may have spread outward and downward by sampling the topsoil and deeper soils, respectively. They do topsoil sampling manually or with small machinery. For deeper samples, heavy machinery trenches the ground until it reaches the sediments below.

Soil sampling and analysis helps in estimating the extent of the contamination that must be excavated. After analysis, the project manager cooperates with the local government and the EPA to design a detailed cleanup plan. Workers meanwhile put secure fencing around the excavation site. When a cleanup plan is ready to be put into action, backhoes and bulldozers begin the heavy work. As they excavate, they often find buried drums leaking hazardous liquids. Hazmat technicians inspect each drum to assess how it should be handled. Earthmoving equipment can remove each drum, but the job requires care to avoid spreading the problem rather than removing it. Workers load the drums and soils onto trucks, which then go to an incineration facility. Dig-and-dump followed by incineration is an efficient way to get rid of hazardous wastes, but it would be even better to altogether eliminate the chances of truck accidents and chemical spills, as well as eliminating incinerator emissions. For this reason today's excavation sites often include new technologies to reduce the amount of soil that must be hauled off the site. In 2007 environmental scientists Alan Seech and James Mueller explained the problem this way in the online publication *Environmental Protection* (URL: www.eponline.com): "... assuming a 25-acre [0.1 km^2] housing development, a total of over 3,000 truckloads of soil would likely need to be hauled out—often through residential areas. Furthermore, an additional 3,000 loads of clean soil would need to be hauled back into the site for backfill. This amount of truck traffic can do a lot to irritate residents of neighboring communities. Finally, the potential for industrial and other types of accidents must be considered a significant downside to this approach."

Cleanup crews treat excavated soils on-site by first separating silts and clays from sand, gravel, and fine particles. Silt and clay adsorb many toxic compounds such as fuels, pesticides, and metals, meaning the material adheres to the soil particles' outer surface. Other soil constituents, however, remain relatively clean, so on-site cleanup includes a soil-washing step in which all the soil flows through a series of sieves and shakers to separate the silts and clays from sand, gravel, and other relatively clean soils. In soil washing, workers then wash the sandy and gravelly soil with detergent and water in large scrubbers. Once these constituents have been completely cleaned, they can be reused. The more difficult-to-clean silts and clays cannot be completely rid of contaminants, and trucks haul most of this material away for disposal. The cleaned soils serve as filler for the excavated area and for later landscaping of the site. Any clean soil left

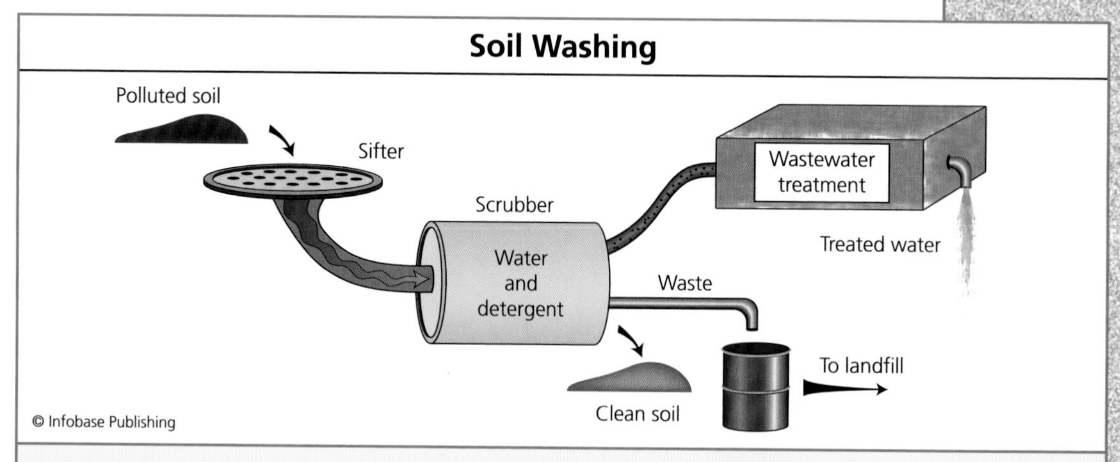

Soil washing removes most industrial chemicals, pesticides, fuels, and metals from excavated soils. Pollutants stick to some soil components, such as silt and clay, more than others, so before the washing step the soil passes through screens that separate fine silts and clays from sands and gravels. The polluted soils then enter the scrubber for cleaning with water and detergent. Once the chemicals have been removed, the cleaned soils can be reused.

over may also be useful as landfill covering. Overall soil washing greatly reduces the amount of soil that must be carried away.

Soil-washing techniques are improving to remove more and more contaminants. One innovation is the use of *solvents* to extract compounds adhering to soil. Solvents work well for removing PCBs, VOCs, halogenated compounds, and petroleum wastes. Two additional enhancements to soil-washing efficiency are: (1) better surfactants, which are detergent-like liquids, and (2) chelating agents, which are chemicals that grasp charged particles. Laboratory scientists also test new surfactant-chelator mixtures for making soil washing even more effective.

Certain hazardous wastes can be chemically treated in the soil before excavation. This on-site treating is called *in situ* treatment. For example some chemicals remove chlorine from chlorinated organic compounds in the soil by a reaction called *dehalogenation,* which reduces the chlorine compounds' toxicity and so reduces potential harm to ecosystems. Dehalogenation has become one of the more useful methods for treating many industrial chemicals and pesticides.

In addition to chemical in situ methods, thermal methods play a role at excavation sites. Portable thermal *desorption* units are about the size of a mobile home and handy because they may be taken directly onto the excavation site. The desorber part of the unit heats contaminated soil to 1,200°F (650°C), and this intense heat causes chemicals to release, or desorb, from the soil. The vaporized chemicals exit the desorber and enter a gas collection device atop the unit, which captures the gases on *activated carbon* beds. (Activated carbon is a blend of carbon granules made to be highly efficient in absorbing pollutants.) Thermal desorption is a good choice for cleaning soils contaminated with fuels, coal tar (a petroleum by-product), wood preservatives, and organic solvents. Metals, which do not vaporize, stream out of the bottom of the unit as ash that is safe for disposal. Current thermal desorbers decontaminate over 20 tons (18 metric tons) of soil per hour and they also clean the difficult silts and clays.

In situ chemical or thermal treatment helps reduce the amounts of contaminants in excavated soils. Thermal desorption works in large cleanup projects. In situ technology, however, requires specialized equipment and extra training, and cleanup managers may feel these things are too expensive. However, as in situ methods become more commonly used, their efficiency increases and costs drop. Most excavation jobs today include some sort of in situ treatment.

Excavating contaminated sediments from under a body of water requires the use of dredging equipment. Most dredging operations for

Thermal Desorption

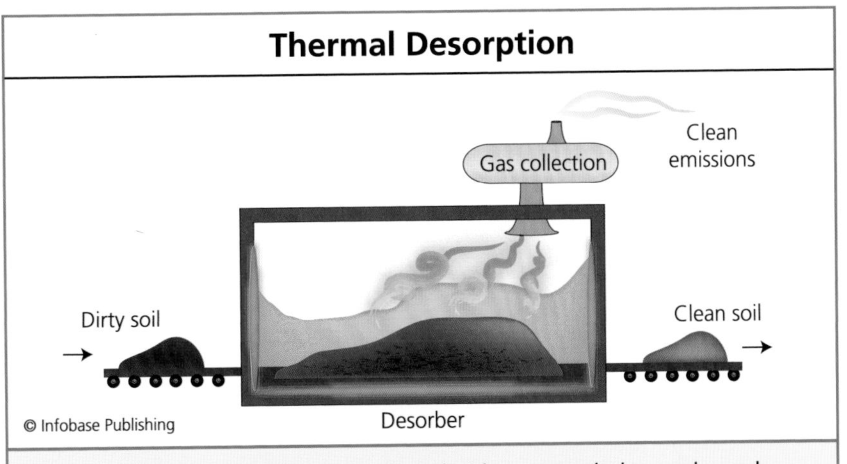

Thermal desorption cleans soils contaminated with compounds that can be made volatile, such as solvents and volatile organic compounds. The intense heat inside the desorber vaporizes the contaminants, which are then collected in a scrubber. This method works well for soils that do not contain high amounts of heavy metals. This method cleans over 20 tons (18 metric tons) of soil per hour, a faster method than soil washing.

removing hazardous wastes take place in rivers and lakes. The cleanup employs the two following types of dredging methods: (1) mechanical dredging with backhoes in shallow flowing waters or places where the contamination has not spread out within the sediments, and (2) hydraulic dredging with pumps for deeper contamination or when the pollutants have formed a slurry (a type of semisolid) within the sediment. The dredging equipment transfers the contaminated materials onto barges for final disposal by incineration or burial.

THE PROS AND CONS OF EXCAVATION

Excavation sites can be messy places in the eyes of their neighbors, especially when large and noisy earthmoving equipment go into action. Soil washing reduces the overall risks of spreading pollution, yet the public may not view this innovation as much safer than the excavations of years past.

Excavation can be the best cleanup choice when more advanced methods are not available, when the site contains a mixture of solid and semisolid wastes, or when an unknown quantity of storage drums lie buried beneath the site. Excavation crews now follow various procedures to keep pollutants from harming the site's neighbors and also to reduce the disruption excavation activities cause to the surroundings. To accomplish these objectives crews install windbreaks or screens to stop dust particles from blowing off the site. During nonworking hours they cover the excavation area with tarps to prevent wind and rain erosion. The Occupational Safety and Health Administration (OSHA) requires that excavation workers also receive protection from the toxic compounds in their midst. Excavation site workers therefore wear safety equipment to protect against inhalation of contaminated dusts and vapors—excavation activities can release pollutants like VOCs into the air. These compounds present an immediate health threat to humans and animals and they also add to the greenhouse effect. (These ubiquitous compounds are discussed in the "Volatile Organic Compounds" sidebar on page 35.)

Excavation sites cannot escape the fact that they are often disruptive, noisy, and dirty places, which might harm their neighborhoods. The irony of excavation is that despite its drawbacks, certain circumstances make excavation the fastest way to rid a place of toxic matter that would otherwise harm a community's health.

(a) Hazardous waste cleanup often involves cleaning up tons of contaminated soil and thousands of gallons of dangerous liquids, shown here at the Hanford Superfund site. *(EPA)*

(b) A hazardous-waste vacuum tanker truck removes contaminated liquids to an incinerator for treatment. *(McRae's Environmental Services)*

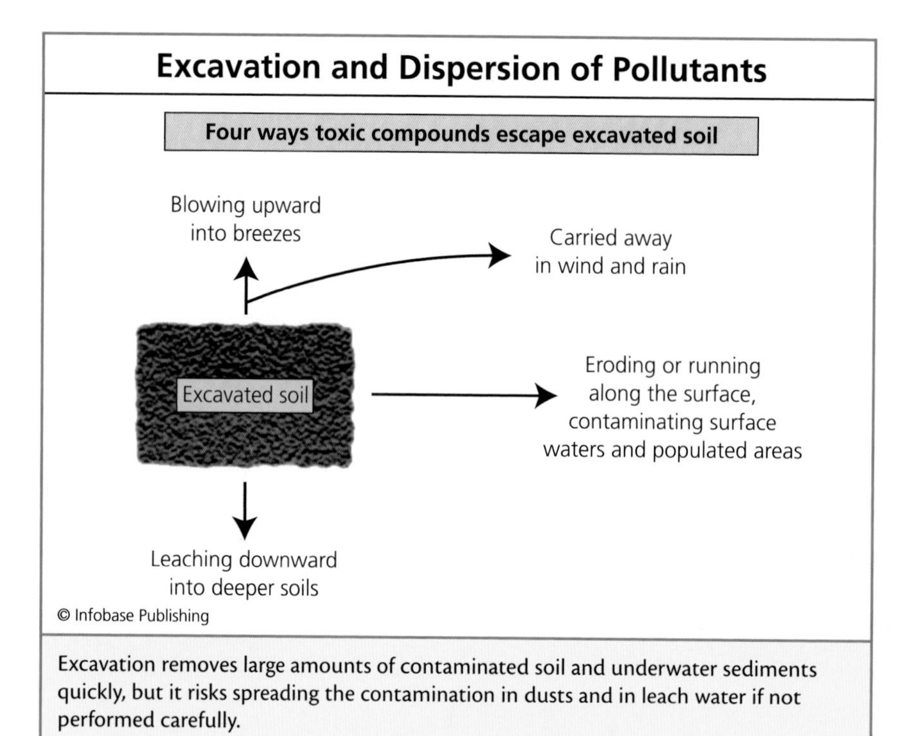

Excavation removes large amounts of contaminated soil and underwater sediments quickly, but it risks spreading the contamination in dusts and in leach water if not performed carefully.

Although excavations on dry land are safer than ever before, aquatic sites are another matter. Dredged soils disperse easily in water and the methods for containing them are not as efficient as on land. In some cases dredging releases pollutants that had been stable in sediments for years; these toxins then enter sensitive ecosystems. Cleanup teams have tried rerouting streams, building dikes, and putting rigid steel sheets around the contamination, but environmentalists fear the overall effect can be worse than the original pollution.

A cleanup project in 2005 provided an example of the problems associated with cleaning up submerged pollution. In that year the EPA discovered that dredging operations along Oregon's Willamette River had released DDT, rocket propellants, acids, and ammonia, plus the metals mercury, lead, zinc, and arsenic. (Sections of the Willamette River had already been designated a Superfund site in 2000, and parts of the river retained significant pollution eight years later.) The cleanup operations had freed substances that had been fairly stable under the river's waters and the materials were carried downstream. Local resident Bill Egan described to the *Oregonian* newspaper seeing "clouds of material leaching into the river" during his boating trips. The silt level in the water became so dense it blocked sunlight from reaching plants and other species in aquatic ecosystems that depended on it. River cleanups are more difficult than most land cleanups, despite better techniques for containment developed in the

Volatile Organic Compounds

A volatile organic compound (VOC) is any carbon-containing chemical that evaporates into the atmosphere. VOCs are of concern in environmental science primarily for the following three reasons:

- They are dangerous to humans and animals when inhaled.
- In sunlight, they mix with nitrogen oxide to form photochemical smog.
- They react with nitrogen oxide to form ozone and nitrogen dioxide.

About 40 percent of the VOCs in the atmosphere come from vehicles burning fuel made of *hydrocarbon* chains—that is, gasoline. During combustion in the engine these hydrocarbons break into smaller volatile chemicals and disperse into the air. In addition to petroleum fuels, solvents, degreasers, and paint thinners also release VOCs; well over 250 VOCs escape into the air each day from numerous sources. The most common VOCs in the air are benzene, acetone, di- and trichloroethylene, vinyl chloride, toluene, and methylene chloride. Long-term, or chronic, exposure to these compounds causes liver and kidney damage and increases the risk of cancer.

The Clean Air Act covers some aspects of VOCs such as requiring gas stations to install vapor recovery systems on gas dispensers to capture volatile compounds. Newer model cars have similar systems, but these are merely small steps toward solving the VOC problem. The EPA has yet to set regulations for the allowable amounts of VOCs that can be made by motor vehicles. In the meantime VOCs continue to drift upward into the Earth's atmosphere.

past few years. Without special care, dredging may create more problems than it solves. As Egan concluded, "It doesn't do any good to clean up any of the site if the uplands are just going to keep polluting the river."

The EPA requires land or water excavation teams to achieve what it calls clean closure before it will designate a site as decontaminated. Clean closure means that the excavated soil has been cleaned and follow-up testing shows that no contaminants remain. The cleanup/treatment methods may differ from site to site, but in the end the EPA decides when the job is finished.

HANDLING EXCAVATED POLLUTANTS

The main benefit of in situ treatment is the reduction of the tons of hazardous soil to be hauled away over roads. Excavated soils that cannot be cleaned need safe handling and safe transport. At the cleanup site, crews first separate the partially cleaned soils according to type: all solids, semisolids, loose soils, compacted soils, and land containing storage drums. These different loads require various transport methods to their treatment or disposal site. In most instances trucks carry hundreds of tons of contaminated soils every day from cleanup sites to landfills, incinerators, or wastewater treatment plants. The EPA and individual states enforce hazmat laws and other safety precautions for this massive level of hazardous material transport.

Certain states require a written plan for handling excavated materials, air monitoring, personal protective equipment, security, and emergency actions. State officials inspect a detailed checklist before each load may be shipped. Other safety precautions used by haulers include: hauling several small loads rather than stockpiling large loads; water-spraying soil piles to reduce dusts until the wastes are loaded onto trucks; tarp coverings; temporarily suspending operations on very windy days; and sloping excavation sites to direct any erosion into the site rather than out to the surroundings.

The truck drivers who transport excavation materials register as commercial waste haulers in their state and maintain their trucks according to federal laws enforced by the U.S. Department of Transportation. In some states, landfill operators prepare a written letter authorizing transfer to the landfill. Overall, today's handling and hauling of excavated materials have made great strides toward safety compared with early dig-and-dump operations. One of the nation's largest excavation projects is highlighted in "Case Study: U.S. Army's Rocky Mountain Arsenal" on page 37.

Case Study: U.S. Army's Rocky Mountain Arsenal

In 1942, early in the United States' involvement in World War II, the U.S. Army claimed 20,000 acres (80.9 km^2) of the Colorado Front Range east of Denver for chemical weapons production. The army named the piece of land the Rocky Mountain Arsenal (RMA) and commenced using the site for the production of chlorine gas, mustard gas, and lewisite, a more lethal mixture composed of chlorine, arsenic, and acetylene. By the end of the war, RMA had produced more than 100,000 tons (90,720 metric tons) of chemicals. The military deactivated RMA after the war, but revived it in the 1950s during the cold war between the United States and

(continues)

Part of the U.S. Army Rocky Mountain Arsenal. Since 1942 this property outside Denver has been either a manufacturer or a depository of dangerous chemicals, nerve gas, and weapons. By 1970, about the time this picture was taken, the army and chemical companies had begun a massive cleanup project. *(U.S. Army)*

(continued)

the Soviet Union. Tons of munitions, incendiary bombs, and the extremely poisonous GB nerve gas were produced. Later, the Shell Chemical Company leased part of RMA's land for pesticide and *herbicide* production, while other parts of the site housed research in rocket fuels. In the 1950s and 1960s the United States' involvement overseas focused on conflicts in Korea, and then Vietnam. The RMA did its part by becoming a major development center for new weapons.

In 1970 the Department of Defense optimistically stated, "... a $14.6 million project for destroying toxic nerve and mustard gas stored at Rocky Mountain Arsenal will begin next fall and be completed in 1973." In 1972 the United Nations held a conference in Stockholm, Sweden, to discuss devastating diseases and deformities that had been emerging worldwide in people and animals due to metal and chemical pollution. Ecology-conscious activists in the United States used this as one of many arguments in favor of new, strong environmental laws. When Congress responded by passing the Clean Air and Clean Water Acts, many eyes fell on the massive waste deposits at RMA.

Throughout the 1970s the U.S. Army carried out the slow removal of its poisons. While the work took place, a steady stream of chemicals trickled out of waste drums yet to be removed. Toxic chemicals percolated through the soil under waste stockpiles and reached the groundwaters below. In the 1980s the army and the Shell Oil Company (the chemical company's new name) entered into a somewhat tense partnership—both sides brought along a large team of lawyers—to carry out decontamination of RMA's groundwaters and soils. RMA would become one of the nation's largest excavation jobs ever undertaken.

The polluters reviewed over 40 different treatment technologies to aid the excavations. Not until 1990 did project leaders settle on one called submerged quench incineration (SQI). SQI heats polluted soils to 1,900°F (1,040°C) to melt the organic constituents, and then cooled, or quenched, these in underground water tanks. Technicians then recover the metals, organic chemicals, and cleaned water and send them either to a disposal site or a facility that reuses them. In some spots, however, SQI could not work. A nightmare of trenches, pits, basins, and pools overflowed with hazardous substances and needed pumping or excavation rather than SQI.

EXTRACTION AND STABILIZATION

Excavation removes polluted soils; *extraction* removes toxic liquids from soils. Pollution extraction can be either active or passive, depending on the energy required for doing the job. Pumping is an active method of clean-

The cleanup of parts of RMA continues to this day. From the beginning of RMA's cleanup to the present, costs have run over $1.3 billion. Notable also, despite the army's predictions in 1970, in 2007 the *Rocky Mountain News* reported that hazardous substances continued flowing into the area's groundwaters. At that time, Colorado's attorney general admitted ". . . groundwater at the Arsenal will not be clean for the foreseeable future."

Shortly after the RMA cleanup had begun decades earlier, archeologists from the National Park Service uncovered spears, knives, and cooking tools dated from 3500 B.C.E. to 1000 C.E. Centuries later, Apache tribes yielded their lands to pioneers who in turn were pushed aside in 1859 during a frenzied but unfruitful Colorado gold rush. In the process 300 native species of animals all but disappeared. Farming and grazing took place until the weapons manufacture began. Today, the EPA oversees restoration of thousands of acres of RMA and its return to natural uses. The Rocky Mountain Arsenal National Wildlife Refuge supports a small herd of bison, and many of the original plains and prairie inhabitants have returned: deer, waterfowl, raptors, prairie dogs, songbirds, and rodents. Despite these successes, RMA's land will never return to its original condition. The 2007 article in the *Rocky Mountain News* reported the area will remain contaminated beyond 2010.

Denver's urban population has expanded into areas surrounding the refuge. Housing and roads block migration corridors, artificial lakes and wetlands have not been planned with wildlife in mind, and invasive trees have taken over large tracts. Naturalists fear that burrowing animals will find pockets of chemicals in water and soils, especially since chemical warfare substances were still being discovered in the refuge as late as 2007. The buildings still standing may release toxic materials into their surroundings when they are finally demolished.

More than 80 percent of the RMA has now been removed from the Superfund list, with the remaining cleanup due in 2010; excavation serves as the main cleanup activity. RMA's excavated wastes go to one of two main disposal routes: landfills, or deep burial in wells, followed by sealing each well. The clock can never be turned back on RMA, and its history includes both encouraging and discouraging news. Nevertheless it offers many lessons on the devastation caused by pollution and the challenges of cleanup.

ing contaminated groundwaters or protecting clean groundwaters located near contamination. In this method, clean water pumped into inflow wells forces contaminated water to the surface through outflow wells, called collection wells. Even *plumes*, underground spreading of contaminants, can be pushed from underground sites by active pumping. If needed, pumps reinject partially treated water repeatedly until the contamination disappears. This method, therefore, is called pump-and-treat.

Pump-and-treat is not foolproof. Small underground plumes can break off from the main contamination area or form when the pumps shut off. Also, the drilling process for assembling a pump-and-treat operation is costly and the entire process can be time-consuming.

A cost-saving variation on pump-and-treat involves the drilling of a single collection well into a clean groundwater source that is threatened by contaminated water nearby. The clean water can then be pumped out and stored permanently aboveground to rescue it from contamination.

In some cases, passive extraction provides the only means for removing organic compounds. For many years cleanup technicians practiced a type of passive remediation by putting bacteria down wells drilled into an organic waste plume. Anaerobic bacteria decomposed the toxic compounds into water and methane gas. The flammable and potentially explosive methane—greenhouse gases had not yet been identified as such—exited through vents built into the Earth. The United States no longer allows this method, called methane venting, but a new and safer technology has replaced it: soil vapor extraction. Soil vapor extraction avoids the buildup of dangerous methane (or other gas) by using vacuums to extract the gases. The extraction apparatus pulls gases upward through tubes in the ground and captures them in activated carbon. The inexpensive carbon granules must be replaced periodically; when the granules no longer work as an absorbent, they are disposed of in a landfill or by incineration. Another advantage, soil vapor extraction causes little disturbance to the land.

Even gentler to the environment than vapor extraction is biological extraction using plants. In this method, called *phytoextraction,* a tree or plant draws metals or organic pollutants into its root system at the same time as it pulls nutrients from the soil. Once phytoextractive plants mature they are harvested, and in this way the contaminants are removed from the soil along with the root systems. *Phytostabilization* provides a variation on this type of passive extraction. In phytostabilization, root systems absorb or simply bind the contaminants to hold them in place in the soil.

Plant scientists seek plant varieties that extract greater amounts of contaminants, and more quickly. These plants are called *hyperaccumulator* plants. Hyperaccumulators that take up metals usually live in areas with naturally high soil metal concentrations. Their roots have evolved the ability to bind large amounts of metal ions, perhaps as a guard against metal toxicity—some plants even absorb radioactive uranium. Unfortunately, hyperaccumulator plants have two disadvantages that must be overcome for them to be useful in contaminant cleanup. First, hyperaccumulator species are rare and they tend to grow in very remote places. Botanists may in the future rely on greenhouses to develop new varieties of hyperaccumulator plants. Plants that grow best will then become test cases in the field. The best choices will probably be varieties that grow well in varied climates, produce fast-growing and large root systems, and accumulate large amounts of contaminants. Second, most of the currently known hyperaccumulators absorb only one metal, so the future of passive extraction may depend on *genetic engineering*, which will develop hyperaccumulators capable of extracting more than one metal. Plant physiologist Leon Kochian, one of the first researchers in hyperaccumulation, summarized the value of hyperaccumulators such as Thlaspi, a plant in the cabbage family. Kochian explained in the June 2000, issue of *Agricultural Research* magazine, "A typical plant may accumulate about 100 parts per million (ppm) zinc and 1 ppm cadmium. Thlaspi can accumulate up to 30,000 ppm zinc and 1,500 ppm cadmium in its shoots, while exhibiting few or no toxicity symptoms. A normal plant can be poisoned with as little as 1,000 ppm of zinc or 20 to 50 ppm of cadmium in its shoots." Molecular biologists are now studying the possibility of putting more than one metal uptake gene into a single plant. The result may be a super-hyperaccumulator plant!

CONCLUSION

Excavation may be the best choice for removing contaminated soils or sediments from polluted areas. Its two main advantages are speed and its capacity for large amounts of materials. But excavated substances must be hauled to disposal sites or treated off-site. To reduce the chance of spilling substances during transport, excavated soils or waters can be treated on-site. Soil washing represents one such on-site treatment in which detergents remove contaminants from soils in a desorber unit. The cleaned soils can then be reused. Contaminated groundwaters usually require pumping and aboveground treatment until they are free of contaminants. Any treatment done at a cleanup site is called in situ treatment.

Excavation operations today use many precautions against spills, leaking, and airborne dusts. Cleanup teams use various methods to prevent blowing dusts and runoff. Haulers who take the leftover contaminated soils to disposal sites must follow strict government regulations on transporting hazardous materials. Underwater excavations are more difficult than land excavations and dispersal of the contaminants remains a problem.

Extraction provides a passive method for drawing contaminants out of the ground without disrupting the land as excavation does. Certain plants absorb pollutants, especially toxic metals, into their roots. Harvesting of the plants thereby removes the contamination. Plant biologists have recently discovered plant varieties with very strong metal extraction ability, called hyperaccumulator plants. Plant remediation, called phytoremediation, of contaminated sites therefore has excellent potential as a follow-up to excavation.

Microbes and Plants for Toxic Cleanup

Bacteria and fungi in the soil degrade a variety of contaminants put there by human activities. Pesticides and herbicides washed off crops by rain enter the soil and leach downward toward *aquifers.* Given time, microorganisms (also called microbes) change the structure of these compounds and so take away their toxic effects. This natural process is *biodegradation,* a type of bioremediation. When plants or trees rather than microbes carry out reactions to degrade compounds in nature, this is termed *phytodegradation.* Both bioremediation and phytoremediation disrupt the land to a far lesser extent than excavation, landfilling, and burial, and they do not expel wastes as incineration does. Despite these advantages, natural processes take a long time to work on contamination compared with mechanical cleanup methods. There may be times when natural remediation is not suitable for pollution cleanup. Given the time, however, natural remediation protects the environment as it cleans it. This is perhaps bioremediation's most important feature.

The most common mechanical methods of pollution cleanup are excavation, incineration, burial, and landfilling. These methods can disrupt ecosystems and spread pollution if they are done carelessly, and communities next to cleanup sites rarely enjoy the sights, sounds, and smells they produce. Natural methods by contrast use biota of the natural environment: soil bacteria and fungi, aquatic organisms, trees, and plant life. Natural remediation restores habitats and ecosystems while it removes pollution. The same thing can certainly not be said about an excavation site or an incineration plant.

Bioremediation was first used in hazardous substance cleanup in Pennsylvania in 1972 by the Sun Oil Company. The company experimented with a mixture of microbes to clean up a petroleum spill, but the full potential of bioremediation had yet to be fulfilled. As the young biotechnology industry grew in the 1980s, it seemed microbes genetically designed to gobble oil spills would become a priceless breakthrough.

In 1989 the *Exxon Valdez* oil tanker ran aground off the coast of Alaska and spilled 257,000 barrels of *crude oil.* This looked to be the first large-scale opportunity for genetically engineered bacteria to clean up an environmental disaster. Bacteria were indeed put to work on the shoreline of Alaska's Prince William Sound to chew up the oil. Cleanup teams did not employ genetically engineered bacteria, however, because the public and political leaders had not yet accepted the idea of bioengineered species released into the wild. The *Exxon Valdez* cleanup depended on the natural bacteria already found in the environment. The *New York Times* quoted Environmental Protection Agency (EPA) Administrator William Reilly as explaining the decision to use bioremediation: "The skimmers and other mechanical means of oil removal used in Prince William Sound at the time of the incident were inadequate. New techniques for beach and water cleanup under a variety of conditions should be developed. Innovative cleanup techniques, for example, microorganisms, should be examined and field tested."

A 1991 *New York Times* article quoted Daniel Abramowicz, environmental scientist at General Electric, as saying that more than 70 percent of the 545 PCB-contaminated samples taken from the Hudson River had shown significant degradation by bacteria found in nature. Within the next few years laboratories began investigating ways to improve on the ability of certain microbes to digest pollutants in the environment.

The use of *genetically modified organisms* (GMOs; also GEMs, for genetically engineered microorganisms) in bioremediation stalled throughout the 1980s and into the 1990s, although there were plenty of opportunities for cleaning up pollution. Natural microbes failed to deliver satisfactory results, despite the successes in Alaska and New York. Bioremediation using GMOs deserved another look. Today bioremediation tasks usually involve combinations of natural species with bioengineered species. Bioremediation generally supplements other cleanup technologies at a single hazardous waste site so that the bioremediation activities take place in a small area within the larger cleanup site.

Like microbial cleanup, phytoremediation first came into use in the 1980s on a large scale when scientists tried it at a nuclear accident near Chernobyl, Ukraine. A variety of plants helped pull contamination from the soils and water near the accident site. In 1990 in the United States, plants cleaned up several nitrogen-contaminated aquifers in New Jersey. Since then phytoremediation cleanup projects have grown in number throughout the United States, and biologists are working to find the best plants for local climates and for specific pollutants.

Microbes and plants possess different characteristics when used for cleaning up the current list of hazardous waste sites. Microbes tend to be used for degrading and detoxifying hazardous chemicals, and plants are used mostly for extracting chemicals from the earth or stabilizing them in soil. This chapter explains new technologies in bioremediation and phytoremediation, how they are used, and ideas for their future use.

REMEDIATION TECHNOLOGY EMERGES

Two main factors in bioremediation planning are the specific pollutants to be treated and the appropriate microbe or plant for cleaning up those materials. The methods for analyzing contamination are discussed in the first chapter. The next step involves selecting the best cleanup tool—microbe or plant—to perform the cleanup.

This selection process, and remediation technology in general, begins with consideration of the type of land to be cleaned—wetland, grassland, industrial property, urban land, etc.—and how it will be used once it is restored. A bioremediation plan also includes an evaluation of the type of soil to be cleaned. This is because soil type influences the plants selected for phytoremediation or the type of microbes appropriate for bioremediation. In either case, soil testing is the first step. Plants and trees—more than microbes—grow only where climate, rainfall, temperature, and soil conditions are favorable to them. Soils vary in moisture, nutrient content, nitrogen levels, pH, clay or sand content, and aeration. Remediation specialists sample the site's soil and measure all of these characteristics before remediation begins. Examples of desired ranges for phytoremediation are shown in the table on page 46.

After defining soil conditions, the remediation plan describes how the land is to be used after it has been cleaned. Phytoremediation may be the perfect choice for land that will be returned to natural habitat for wildlife.

OPTIMAL CONDITIONS FOR PHYTOREMEDIATION	
ENVIRONMENTAL CONDITIONS	OPTIMAL RANGE
available soil moisture	25% to 85% water-holding capacity
reduction-oxidation potential	Eh greater than 50 millivolts
oxygen	greater than 0.2 mg/L dissolved oxygen
nutrients	carbon:nitrogen:phosphorus molar ratio of about 120:10:1 (mg per kg of soil)
pH	5.5 to 8.5
temperature	60°F to 110°F (15°C to 45°C)

On the other hand microbes might be the best choice for cleaning oil-slicked rocks along a shoreline.

The chemical form of the contaminant also affects whether plants or microbes are the best choice. Plants excel at extracting metals from soils or sediments but they do not always work on organic contaminants. Many bacteria thrive on organic contaminants in soil or water, but metals are toxic to them. Unfortunately, the worst contaminated sites contain many different chemicals and a nightmarish combination of breakdown products. This is one reason why natural remediation methods work best when combined with mechanical cleanup technologies.

PRIORITY POLLUTANTS

Priority pollutants are the substances that should be removed first from a hazardous waste site. Military bases and large manufacturing sites tend to have mixtures of these priority pollutants. Examples of priority pollutants are wastes in danger of exploding or igniting, or very toxic chemicals known to persist in the environment, such as synthetic pesticides. As mentioned in chapter 1, persistent chemicals present the greatest long-term threat to ecosystems. Persistence of a chemical relates to the number of years after its release that instruments are able to detect it in the environment. Obviously, chemists can measure only to the lowest level that their analytical equipment can detect. Chemicals therefore may persist in the environment in minuscule amounts for many years longer than scientists realize.

Persistent compounds resist being degraded by the following factors in addition to microbial action: the Sun, extreme temperatures, water, and other chemicals. All of the persistent compounds listed in the table below are pesticides except for dioxins and furans. Dioxins and furans are industrial chemicals used in the manufacture of various plastics and are released when plastic burns. Dioxins and furans also come from the burning of diesel fuel, home heating fuel, wood, and tobacco, as well as electric power plants and iron and steel manufacturing.

Persistence of Organic Compounds in Soil

Compound	Persistence in Years
DDT	21
dieldrin	21
chlordane	16
heptachlor	16
toxaphene	16
dioxins	8
furans	8
paraquat	6
picloram	5
chlorfenvinphos	4
trifluralin	3

Relative Biodegradability

Increasing biodegradability

Simple hydrocarbons, petroleum fuels
Aromatic hydrocarbons
Alcohols, esters
Nitrobenzenes, ethers
Chlorinated hydrocarbons
Pesticides

Microbes decompose most organic compounds in the environment. Some pollutants stay in the environment far longer than is desirable because microbes do not have enzyme systems to efficiently degrade these compounds.

Compounds synthesized in laboratories persist in the environment because microbes in soil or water do not have enzymes to break them down. These substances are not normally found in nature and are therefore completely foreign to living things. Any compound made outside of nature is called an *anthropogenic* or *xenobiotic* compound. Many pesticides and herbicides are xenobiotic. Microbes have been developed to degrade xenobiotic compounds in laboratory experiments, but these species are not as effective in the environment. This is because in nature microbes metabolize nutrients by the least energy-demanding means. In nature bacteria usually find compounds easier to metabolize than xenobiotic structures, and when this happens the microbe ignores the synthetic version. As a consequence the chemical compound remains intact in the environment.

Even some natural compounds persist in soil for centuries. Compounds found in fossils, sedimentary rocks, and coal remain intact for hundreds of thousands of years. For example, resins in amber last for over 100,000 years. Even some microbes outlast humans by a wide margin. Bacterial spores and various fungi have been isolated from archeological discoveries—*Bacillus* spores have been found in Egyptian mummies—many centuries old. The hardy nature of microbes makes them useful in bioremediation.

DETOXIFICATION BIOLOGY

Microbes often see compounds that are toxic to animals as an appetizing blend of carbon, cellular building blocks, and substances needed for energy production. When microbes degrade hazardous compounds to safe end products, they are said to detoxify the compounds. Microbial cells use carbon, nitrogen, phosphorus, and other elements in their *metabolism,* and as a fortunate result, toxic compounds become less toxic. This detoxification takes place because a group of microbes either break apart the toxic compound or rearrange it into a new compound. This process of detoxification is also sometimes referred to as neutralization of the toxic compound.

Microbes at the aerated surface of the earth or those living in lightless, airless sediments both detoxify hazardous chemicals. Inside the microbial cell, electrons pass through a chain of reactions to produce energy in the form of the compound adenosine triphosphate (ATP), which powers the cellular machine. Aerobic species living at the Earth's surface use oxygen as part of this *electron transport chain.* Anaerobic species in deep sediments use nitrates, sulfates, or metals instead of oxygen in energy generation. Some microbes use the hazardous compound itself in their

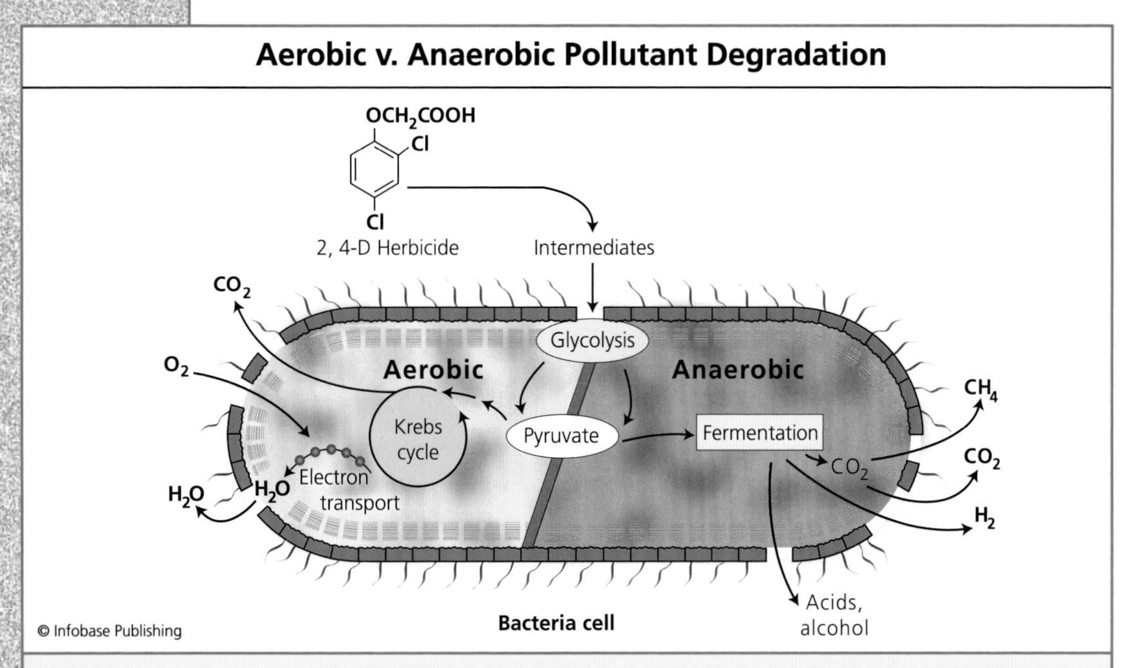

Aerobic and anaerobic bacteria use glycolysis for energy production. Bacteria first degrade contaminants into smaller compounds that feed into the glycolysis pathway, then aerobes and anaerobes use different methods for making energy. Aerobes live in shallow, aerated soils and waters; anaerobes thrive in deep sediments and stagnant waters.

energy-producing systems or as a carbon source. The microbe therefore gets a benefit from the hazardous chemical. At the same time the environment benefits because a hazardous waste has been converted into a harmless end product.

Some microbes degrade xenobiotic compounds completely to carbon dioxide and water. Others carry out only partial degradation. Complete or partial breakdown of any compound for the cell's use is called *catabolism*. In either case microbes degrade compounds for one reason: energy production. Microbes will degrade pollutants if they possess the proper enzymes and if the steps generate energy for the cell. They are much less likely to spend energy to degrade a pollutant.

The more dissimilar a xenobiotic structure is from naturally occurring compounds, the less likely degradation will go completely to carbon dioxide, water, and other small molecules. But pollutants have been in the environment for a long time and microbes have learned to team up to degrade unusual structures. Oftentimes a microbe picks up catabolism where another microbial species leaves off. When more than one microbe degrades a compound, the process is called *cometabolism*. Cometabolism takes place in natural microbial communities unseen by people. Every day microorganisms detoxify unknown amounts of hazardous compounds. Detoxification biology therefore plays an important role in pollution cleanup. It may be especially useful in eliminating hazardous compounds that have been missed by mechanical cleanup or compounds that go undetected in the environment.

Microbes detoxify today's detergent formulas that escape into the environment, but this was not always the case. In the 1960s chemical companies offered their customers new synthetic detergents that worked better than any previous cleaning products. Rinse waters from car washes and household washing machines carried these detergents into surface waters. Almost immediately microbial communities in detergent-polluted ecosystems succumbed under a heavy xenobiotic foam layer. The microbes were unable to chew up the detergents' branched-chain hydrocarbons. Other living things in polluted ecosystems faltered as well. As the problem became more apparent—lakes and streams covered in froth were clues—detergent manufacturers changed the chemical structure of their products to eliminate the branching from the main molecule. These more linear chains mimicked natural compounds; nature's bacteria, algae, and fungi began at once to degrade them. The revised detergents

Detergent pollution of surface waters. After chemical companies introduced synthetic detergents, these compounds and their phosphate chemical groups caused ecosystem destruction because natural systems could not degrade the synthetic substances. Many industries now strive to create biodegradable compounds that decompose in the environment. *(U.S. Fish and Wildlife Service)*

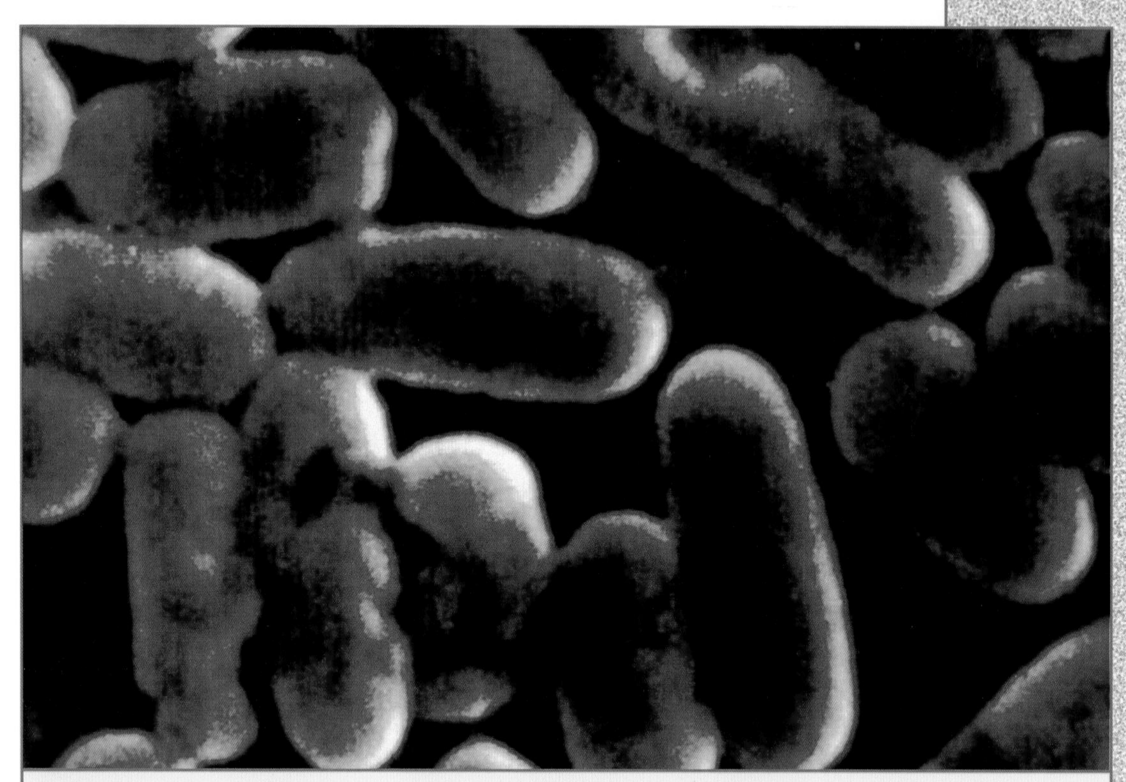

Pseudomonas aeruginosa bacteria are common inhabitants of natural waters. Environmental science increasingly relies on the natural processes carried out by this microbe and others in water and soil to degrade contaminants. *(Pall Corporation)*

sold by the end of the 1960s introduced a new concept: biodegradability. The new chemical structures allowed certain products to degrade more easily in nature.

Microbial detoxification of hazardous wastes is now known to take place by two related modes of action: biodegradation and *bioconversion.* Biodegradation is the complete breakdown of a dangerous compound by a single microbe or more than one microbe by cometabolism. Bioconversion is the transformation of a compound into a less toxic form. Microbes bioconvert compounds by removing part of the chemical structure or adding something new to it. (When plants carry out this same process, it is called *phytotransformation.*) Bioconversion may be a simple step such as the removal of a chlorine molecule. Other types of microbes may detoxify compounds by adding a molecule to them. For example, soil bacteria detoxify phenol compounds by adding a methyl group to phenol's carbon ring. The table below gives examples of detoxification carried out by bacteria in the presence of oxygen.

Detoxification of Pollutants		
Type of Pollutant by Chemical Group	**Chemical Description**	**Aerobic Detoxification Pathways**
alkane	long straight-chain compound, mainly carbon and hydrogen	oxidize terminal carbon to form fatty acids
alkene	similar to alkane but with one or more double carbon bonds in the chain	oxidize carbons to form alcohols
alicyclic hydrocarbon	long-chain carbon and hydrogen containing a cyclic group	open ring structure to form oxygenated hydrocarbons
aromatic hydrocarbon	cyclic compound made of carbon and hydrogen	open ring with oxygen to form precursors for the tricarboxylic acid cycle (Krebs cycle)
halogenated compounds	any compound containing a halogen side group, such as chlorine or fluorine	remove halogen (dehalogenation) and add water molecule to form fatty acids
heterocyclic compounds	cyclic compound containing carbon, hydrogen, and one or more other element	ring cleavage, dehalogenation, or linking with other molecules
metals: mercury and arsenic	elemental metal or metal in a compound with other elements	addition of one or more methyl groups (*methylation*)

NATURAL DEGRADATION

People have depended on natural systems to degrade wastes for centuries. Wastes discarded on the outskirts of settlements eventually turned into harmless material under the forces of Sun, rain, and time. Ship crews have for centuries counted on the ocean's salt and creatures to detoxify the wastes they poured overboard. Before Congress enacted today's environmental laws, manufacturers disgorged thousands of gallons of waste into rivers with the hope that microbes would make them vanish (in fact, microbes did manage to destroy a large amount of the wastes in intentional or inadvertent releases). Today pollution control measures purposely include the activities of bacteria and fungi to degrade toxic wastes.

Degradation of dangerous compounds polluting the environment depends on enzymes made by living things in nature. Some of this natural activity begins to work immediately at the hazardous waste site. Any natural process that carries out reactions such as waste degradation is called *intrinsic activity.* Natural degradation is similarly called *intrinsic bioremediation* (also *natural attenuation*) and it provides one of the safest ways to return a polluted area to a clean space for wildlife or humans. But intrinsic bioremediation has two disadvantages: It works slowly, and different plants and microbes offer a wide range of effectiveness. Weather, erosion, and flooding also alter intrinsic bioremediation. Finally, even proponents of natural systems admit that they do not work on all hazardous substances, and some toxic compounds persist in the environment regardless of the biological or physical factors that break down other compounds.

Wetlands provide perhaps the best conditions for naturally detoxifying the pollutants entering them in rain and runoff. Plants, grasses, and their root systems filter impurities from the water. At the same time, the slow flow of water allows microbes more time to carry out their activities. Flowing systems contain a specialized collection of microorganisms called *biofilm.* Biofilms contain mixtures of bacteria, algae, fungi, and large polymer compounds, all securely attached to a surface such as a submerged plant or rock. As biofilm draws nutrients out of water, it also binds metals and captures organic compounds. In this way natural biofilms remove the metals mercury, lead, cadmium, chromium, copper, and zinc from the polluted waters that flow into wetlands.

Treatment plants in Germany, the Netherlands, and Japan construct biofilms on solid surfaces inside bioreactors to remove contaminants from

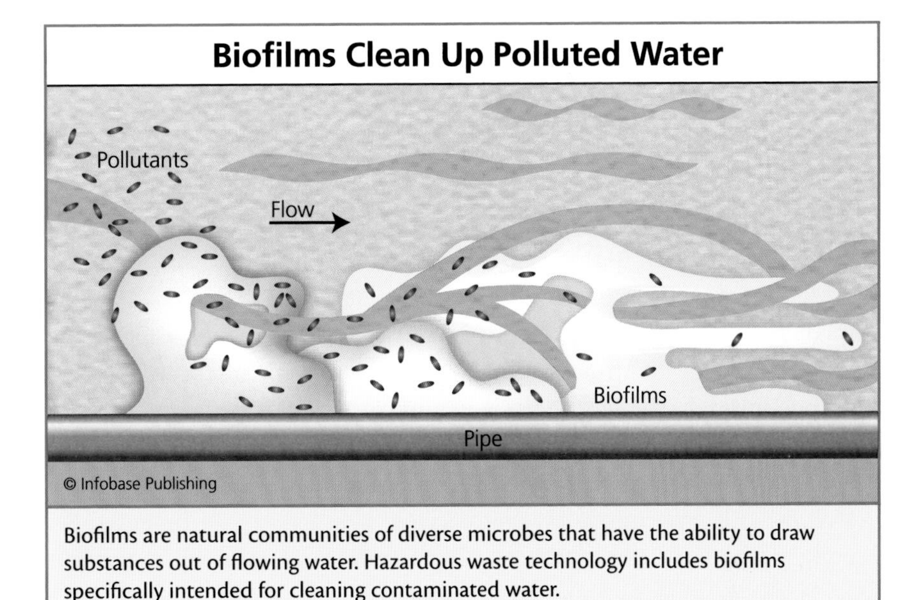

Biofilms are natural communities of diverse microbes that have the ability to draw substances out of flowing water. Hazardous waste technology includes biofilms specifically intended for cleaning contaminated water.

water. In these reactors, biofilm attaches readily to inert surfaces such as diatomaceous earth, glass pieces, polyurethane foam, and polyacrylamide beads. Biofilms in nature may contain hundreds of microbial species, but bioreactor films contain mostly *Pseudomonas, Arthrobacter,* and *Brevibacterium* bacteria and the fungus *Trichoderma.* The metabolism of these and other microbes collectively degrade the following organic compounds: phenols, pentachlor phenols (PCPs), anilines, and acrylonitrile.

Environmental science includes the study of biofilm and other types of microbial communities. In order to make natural systems work faster and more efficiently, laboratories develop genetically engineered microorganisms (GMOs) that contain enzymes that degrade organic pollutants. Many questions still remain unanswered about the effects of GMOs on the rest of the environment, so science and the public wonder about the overall merits of GMOs. A non-GMO approach relies on the addition of growth factors directly into the earth to enhance intrinsic bioremediation. Small additions of nitrogen and phosphorus compounds to the soil usually improve the growth of natural microbes. Sometimes the soil merely needs aeration by tilling or piping air down tubes to help aerobic bacteria grow. The Raymond Method, invented in 1972 by the Sun Oil Company, provides an example of aiding intrinsic bioremediation by adding nutrients to the soil. Sun Oil had been attempting to clean up a hazardous spill that had leached into groundwaters, so company technicians added nutrients to surface waters, and then pumped the water into the contaminated groundwater below. This Raymond Method therefore treats underground contamination by both intrinsic activities and mechanical methods.

Sun Oil Company's experiment showed the potential of combining natural systems with designed apparatus. Not only would this approach be safe to people and the environment, but it would provide a speedier remediation. Today almost all wastewater treatment plants receive contaminated water, which passes through a sequence of filters to clarify it and then into tanks where bacteria degrade organic matter. This is a very simple use of natural-mechanical remediation.

Biopiles and *bioventing* are two newer means of combining natural activities with environmental engineering. The U.S. Navy invented the biopile method to clean excavated soils by mixing the soil with microbes

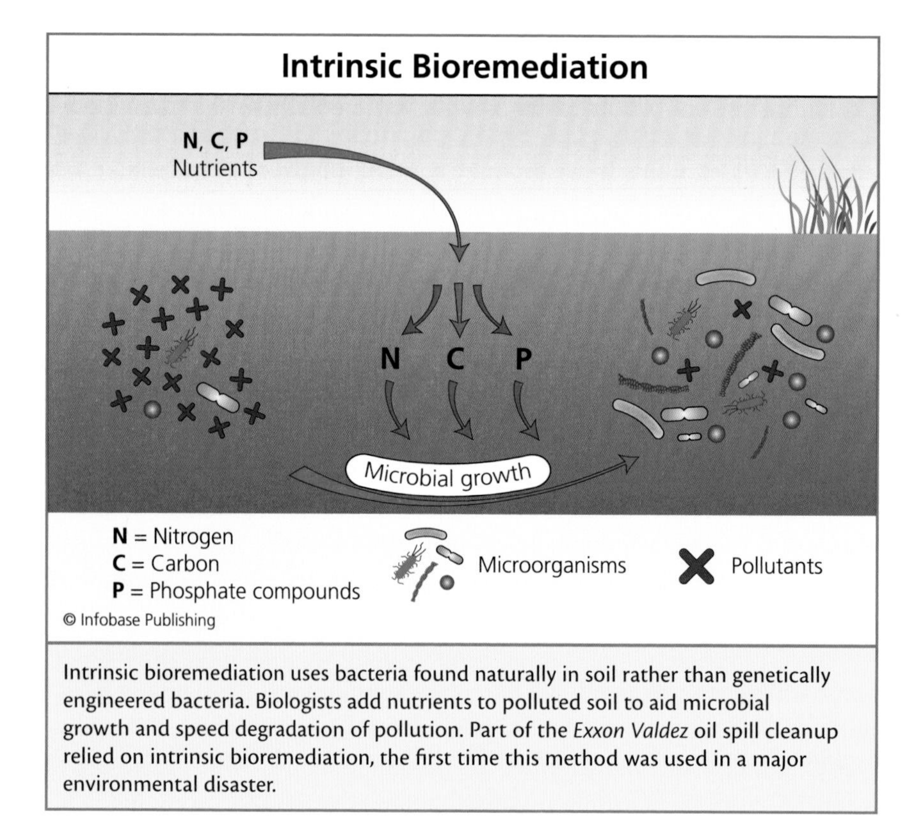

Intrinsic bioremediation uses bacteria found naturally in soil rather than genetically engineered bacteria. Biologists add nutrients to polluted soil to aid microbial growth and speed degradation of pollution. Part of the *Exxon Valdez* oil spill cleanup relied on intrinsic bioremediation, the first time this method was used in a major environmental disaster.

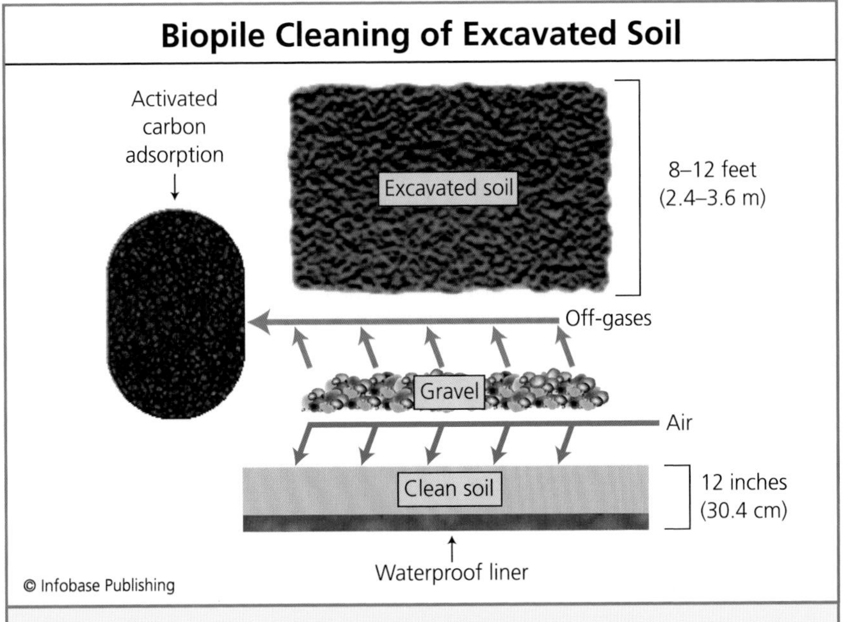

The U.S. Navy has used the biopile method as an inexpensive way to degrade waste stockpiles. The biopile works like a large-scale compost pile: Aerobic bacteria degrade organic contaminants with the aid from added nutrients, water, and aeration

and nutrients, all sitting atop an impermeable barrier laid over a hazardous waste area. Technicians adjust biopile moisture levels, temperature, and aeration to favor microbial reactions and increase degradation. Bioventing, by contrast, is a method of cleaning up hazardous substances underground. In this method, the cleanup crew sends a steady stream of air into wells drilled into underground pollution. The air helps the bacteria in the earth grow faster, and as they grow they degrade organic compounds within the soil. Cleanup techniques that rely on natural systems to remove contamination have shown promise in both routine cleanups and in environmental accidents, as discussed in the "Case Study: Bioaugmentation—the *Exxon Valdez* Oil Spill" on page 58.

BIOENGINEERED MICROBES

Aerobic and anaerobic bacteria, actinomycetes (a type of filamentous bacteria), algae, and fungi grow well in nature and have now been used for remediating polluted soils and water. Of the bacteria, *Bacillus* and

Pseudomonas offer the best attributes for bioremediation. (Other species commonly used in bioremediation are listed in appendix C.) One species of *Bacillus* is highlighted in the sidebar "*Bacillus thuringiensis*" on page 63. *Bacillus* and *Pseudomonas* are both comfortable in a variety of habitats, they grow well in laboratory conditions, and they adapt readily to new environments. *Bacillus* is prevalent in soil, so it is a useful bioremediation

CASE STUDY: BIOAUGMENTATION—THE *EXXON VALDEZ* OIL SPILL

In March 1989, the *Exxon Valdez* oil tanker ran aground on Bligh Reef in Alaska's Prince William Sound. Over 11 million gallons (42 million l) of crude oil gushed through the ship's damaged hull and flowed into the sound's waters. The crude oil, being less dense than seawater, stayed on the surface where the vigorous March waves and winds stirred it into a water-oil mixture called mousse. Normally cleanup crews would remove an oil slick by igniting it and letting the oil burn off, but the high water content of the mousse prevented this from working. At least 4 million gallons (15 million l) of foamy oil drifted onto the shoreline and coated 1,300 miles (2,092 km) of habitat used by invertebrates, birds, otters, sea lions, and important predator species.

The spill taught environmentalists two critical lessons. First, even a medium-sized spill—the *Exxon Valdez* spill ranks only 35th among all oil tanker spills since 1967—could devastate a pristine coastal ecosystem. More than 260,000 waterfowl and raptors, 2,000 sea otters, and 300 harbor seals, plus untold tide pool creatures, shellfish, and salmon populations lost their lives on the beaches and in the waters of Prince William Sound. Many species have not fully recovered their numbers. But a second and surprising lesson emerged: Ecosystems could make significant recoveries with the help of natural microbial systems. In other words bioremediation already took place in nature.

The Alaska spill seemed a prime opportunity for large-scale bioremediation. Bioengineered bacteria would team with mechanical excavation and manual cleanup to restore as much of the coast as possible. Microbiologists knew that hydrocarbons in oil serve as an easily digestible nutrient source for many microbes. Indeed, shortly after the oil entered the sound, marine bacteria burst into a rapid period of growth called a *bloom* due to the generous new supply of carbon compounds. As the bloom grew, the microbes depleted nutrients such as nitrogen and phosphorus from the water and the microbial growth began to slow down as the nutrients disappeared. The large bloom removed some oil, but much remained on the seas and washed onto the shore where the nutrient-starved microbes could not degrade it.

Cleanup crews collected excess oil off the water's surface with booms and washed oily rocks on land with high-powered jet sprays. Volunteers rescued shorebirds and cleaned them tool for cleaning up the soil, whereas *Pseudomonas* is a common water microbe and serves as a logical choice for aquatic sites. A Superfund site in New York received treatment in the early 2000s from a little-known bacterial genus named *Dehalococcoides*. Remediation workers pumped a suspension of the microbe into toluene-contaminated ground. In a *New York Times* article in 2003 about the cleanup, the project's legal adviser,

by hand with mild soaps. Scientists at the site meanwhile devised ways to keep the microbial action going in the hidden oil-damaged places. They tried a new method of bioremediation called *bioaugmentation*. Though the microbes had ample carbon—thanks to enormous amounts of oil—they needed additional nutrients. The scientists experimented for weeks with fertilizers rich in nitrogen, phosphorus, and trace nutrients, until they came upon a promising formula. John H. Skinner of the EPA described it in a *New York Times* article that year: "Essentially, all the microorganisms needed to degrade the oil are already on the beaches, but they need a helping hand in the form of phosphorus and nitrogen nutrients. We plan to give them the needed nutrients by feeding them fertilizer. Laboratory tests have already shown that the technique works, but this will be the first large-scale test ever conducted against a real oil spill."

The material had been formulated to stick to the shore's rocks and pebbles and release nutrients slowly into the surroundings. The slow and steady nutrient supply augmented the metabolism of native microbes and the oil again began to degrade away. In some areas oil disappeared five times faster than in places not getting the nutrients. This method of working with native species goes by several other names: biostimulation, natural attenuation, or *bioattentuation*. They are all terms for intrinsic bioremediation, which is the use of species found naturally in the environment to carry out a waste cleanup. Bioaugmentation still needs improvement, but it has the advantage of relying on natural processes and avoids any unknown problems that might come from using GMOs.

The *Exxon Valdez* accident still affects Alaska's coast. Though bioaugmentation has been credited with speeding the natural breakdown of oil from 20 years to three, bits of oil remain in the sound and hundreds of miles away from the accident site. The *Exxon Valdez* has been repaired and continues traveling the globe with a new name, the *Sea River Mediterranean*, though it no longer enters Prince William Sound. The *Exxon Valdez* spill offered valuable lessons in bioremediation and also showed the long-term harm caused by hazardous waste spills.

Peter Herzberg, said, "The bugs flow where the contamination is, because they're following the water. And you're getting to places where you couldn't get to mechanically with a pump-and-treat system."

Microbes recovered directly from the environment are called wild strains. Environmental scientists now use molecular biology techniques to add new traits to these strains, turning them into GMOs equipped for bioremediation tasks. For instance, in 2004 Daniel van der Lelie of the Brookhaven National Laboratory in New York inserted the mechanism for breaking down toluene from one species of bacteria to another species. Many such GMOs, nicknamed *superbugs,* are superior to wild strains in degrading industry's most persistent chemicals. Opportunities are almost limitless for developing new strains that attack today's most troublesome pollutants. Dr. van der Lelie took the experiment a step further by inserting the toluene genes from bacteria into poplar plants to improve their capacity for cleaning toluene out of groundwaters.

Inventing a new superbug begins by taking a sample of bacteria from a contaminated site. A microbiologist tests individual strains from this sample for their ability to use a pollutant as a nutrient source for growth. Usually pollutant-degrading strains make up a very small proportion of the large number of microbes that grow in nature because their special activity derives from a mutation. The desirable mutants contain a rare gene that directs the cells to produce a pollutant-degrading enzyme system. The process of finding such a valuable gene is called *gene selection.* Once such a gene is identified, the microbiologist inserts it into the deoxyribonucleic acid (DNA) of *Bacillus* or *Pseudomonas* or almost any other microbe suited for the environment. Technicians grow the new bioengineered strain in a laboratory until it produces billions of superbug clones, which are identical new cells made from the parent cell. The technicians then return to the contaminated site and inoculate the soil with bioremediation superbugs.

Gene transfer is an innovative technique similar to the creation of superbugs described here. In gene transfer, technicians put DNA containing the degradation gene or genes directly into contaminated soil. Native microorganisms in the soil then take the new DNA into their cells and so acquire the ability to degrade pollutants. Gene transfer currently shows the greatest potential for success in the use of natural soil bacteria and biofilms for oil spill cleanups.

Not surprisingly, inventing a superbug through bioengineering or gene transfer is not as simple as described here. In November 2006, *Microbe* magazine quoted microbiologist Chris MacKenzie at Houston's University of Texas Health Science Center in describing the situation of finding new microbial genes. "Genome sequencing is a little bit like a voyage of discovery," he said. "You are pretty sure before you set off that you will discover something new, but yet you have little idea as to what that new thing will be." Some of the hurdles to overcome in the development of superbugs are the following. First, the new bioengineered strain may grow well in a laboratory but cannot survive in the environment. Second, soil has a strong capacity to bind organic matter such as bacteria. Superbugs poured onto a contaminated area may stay near the surface and fail to migrate to the contaminants farther away. Perhaps the greatest hurdle GMO technology must overcome is public resistance to unnatural species in the environment. Scientists have debated the difficult question of which problem is of greatest concern to society: the effect bioengineering has on healthy ecosystems or the continued damage inflicted on ecosystems by hazardous wastes.

On-site bioreactors afford a compromise between both sides of the GMO argument. A bioreactor is a vessel that can hold up to hundreds of gallons of liquid and provides the conditions favored by bacteria. In this methodology, superbugs growing inside a bioreactor degrade hazardous compounds as a steady stream of contaminated soil or water enters the vessel. Clean restored soil or water exits the bioreactor while new batches of contaminated matter enter it. Bioreactors' biggest advantage lies in the fact that GMO bioremediation takes place without releasing the engineered microbe into the environment.

Extremophile technology also holds promise because this technology uses bacteria that can be thought of as natural superbugs. Extremophiles make up a diverse group of bacteria that thrive in locations on Earth that are highly toxic or inhospitable to most every other form of life. They have been found in acidic (pH 2.5) mine tailings, sediments laden with heavy metals, DDT-soaked soils, and petroleum-coated shorelines. Some extremophiles have mechanisms enabling them to live in high concentrations of mercury, lead, chromium, cadmium, copper, manganese, or selenium. These unique microbes resist damage from toxic chemicals

Bacteria called extremophiles often live in polluted places. This scientist recovers bacteria from rock 2.2 miles (3.5 km) underground at a South African gold mine. These endolithic bacteria (bacteria that grow in rock) may possess enzymes useful for acting on toxic metals. *(Duane Moser)*

because they have evolved to possess specialized enzymes or protective cellular structures. Many of these organisms actually prefer to use pesticides, organic solvents, acids, or toxic metals in their metabolism. In addition certain extremophiles work well in very hot or very cold environments where other bacteria suffer. Extremophiles may well be the next natural and ultimate weapon in attacking hazardous waste sites. Authors Michael Madigan and Barry Marrs described microbiology's knowledge of extremophiles in 1997—a small number of extremophiles had already been discovered some 40 years earlier—in a *Scientific American* article. "Amazingly, the organisms do not only tolerate their lot; they do best in their punishing habitats and, in many cases, require one or more extremes in order to reproduce at all."

BACILLUS THURINGIENSIS

One of the first bacteria put into service in the environment was *Bacillus thuringiensis*, nicknamed *Bt*. Like all *Bacillus* species, Bt withstands harsh conditions by converting into an almost indestructible endospore structure, which withstands extremes in temperature, drying, chemicals, and exposure to ultraviolet light. Bt in its hardy endospore form may be thought of as science's first true superbug.

In 1911 biologists discovered rod-shaped bacteria growing in the land beneath Thuringia, a province in central Germany. In laboratory experiments, the cells made peculiar crystals that seemed to possess some sort of biological activity. Scientists soon identified the species as Bt and found that the organism's crystal secretions killed moths and other insects. Before long, farmers began spraying Bt suspensions onto their fields to eliminate pests and increase crop yields. The endospore form of Bt offered two conveniences: Suspensions were easy to store until it came time to spray the crops, and the endospores resisted harsh weather conditions in the fields. As the temperature and moisture conditions improved, the Bt endospores on the sprayed plants awoke as healthy reproducing bacterial cells in a process called germination. The germinated Bt cells produced the deadly toxin that insect pests had never before confronted. In this way Bt became the first biopesticide produced by people specifically to help in growing crops and garden plants.

Seventy years passed before biotechnology emerged, but one of the new industry's first forays into recombinant DNA technology involved the transfer of the Bt pesticide crystal gene into plant chromosomes. The new bioengineered plants produced their own pesticide, allowing farmers to reduce the time and expense of spraying their crops. Equally important, biotechnology had found a way to reduce the need for xenobiotic chemicals in agriculture, and so less chemical pesticide entered the environment.

Biotechnology and genetics laboratories now produce superior phytoremediation plants by methods similar to Bt bioengineering. Researchers in many countries have developed plants for cleaning up the following contaminants: detergents, heavy metals, arsenic, selenium, petroleum

(continues)

(continued)

hydrocarbons, and organic compounds. India especially has put intense efforts into phytoremediation research for the purpose of breeding a variety of plants for specific cleanups. In *EnviroNews*, the newsletter of the International Society of Environmental Botanists, authors U. N. Rai and Amit Pal in 1999 provided one example when they wrote, "In India, aquatic vascular plants . . . have been used to treat chromium contaminated effluent and sludge from leather tanning industries."

Root systems then absorb the freed chemicals and thereby prevent them from migrating into groundwaters. It all began with one of the most common bacteria found in soil.

PHYTOREMEDIATION— USING PLANTS FOR TOXIC CLEANUP

In phytoremediation, plants rather than microbes act as the main mode of cleaning up a contaminated site. Plants can remediate both organic compound and metal contaminations, but they are usually used for removing toxic metals from soil or aquatic sediments. When plants remove pollutants completely out of a contaminated site, the process is called phytoextraction. Some of the prominent metal-extracting plants used in phytoextraction today are listed in the table on page 65.

Plant or tree roots work by one of two types of phytoremediation: phytostabilization or phytoextraction. In phytostabilization, pollution is immobilized by vegetation, but not necessarily removed. Phytoextraction by contrast removes hazardous chemicals. Either type of phytoremediation for cleaning up hazardous materials offers advantages to the environment. In addition to removing dangerous chemicals, phytoremediation increases the overall amount of vegetation in an area, which draws the greenhouse gas carbon dioxide out of the atmosphere. Other advantages of phytoremediation are the following:

- easy to learn and implement
- costs little in dollars and energy input

- produces no pollution while it consumes carbon dioxide
- reduces the burden put on landfills and incinerators

Even with its advantages, phytoremediation has some drawbacks. First, plants and trees require months or years to detoxify an area compared with quicker excavation methods. Second, stabilization or extraction of hazardous materials occurs only as deep as the root system grows. Vegetation can prevent chemicals from leaching from topsoils to deeper soils and groundwaters but they cannot remove pollution they cannot reach. Third, plants used for extraction become a hazardous waste that must be handled properly. Finally, hazards that have been immobilized in roots by phytostabilization are not removed from the environment. Insects, birds, or mammals feeding on these plants become exposed to toxic chemicals. Some of these disadvantages can be overcome by using phytoremediation in combination with other cleanup methods.

Phytoremediation advances will soon come from the fields of genetic engineering and plant physiology. Molecular botanists use genetic engineering to put genes into plants, giving them better capacity to remove contaminants from soil. Science may soon create superplants, similar to microbial superbugs.

Examples of Phytoextraction Plants

Plant Family	Examples, Common Names	Metal	Region
Asteraceae	Spanish needles	lead	China
Brassicaceae	alyssum, wild radish	nickel zinc	Southeast Asia Northern Europe
Euphorbiaceae	gooseberry, castor bean, poinsettia	nickel	Hawaii
Lamiaceae	mint, sage, rosemary	copper	Central Africa
Scrophulariaceae	snapdragon, foxglove	cobalt	Central Africa

In the online National Geographic News (URL: news.nationalgeographic.com), microbiologist Guy Lanza of the University of Massachusetts described the overall objective of bioengineering plants. He said, "We're trying to isolate, and hopefully manipulate, the plant systems to do more of what we want them to do. In this case to remove these toxic metals."

Even native plants that have not been genetically transformed make hazardous chemicals less dangerous. Highly efficient hyperaccumulator plants, introduced in the previous chapter, collect more chemicals from soil or water than other varieties. Experts in plant physiology are finding compounds called *chelators* to further help plants extract contaminants. Chelators are compounds that bind metals in a clawlike structure, and so helps the roots absorb metals that would otherwise remain bound to soil particles. The chelator compound ethylenediaminetetraacetic acid (EDTA) serves as a common cleanup substance for this purpose. The aboveground parts of plants also help the environment in two ways: Low-growing vegetation reduces water and wind erosion, and root systems prevent *soil creep,* a slow migration of soil down a slope.

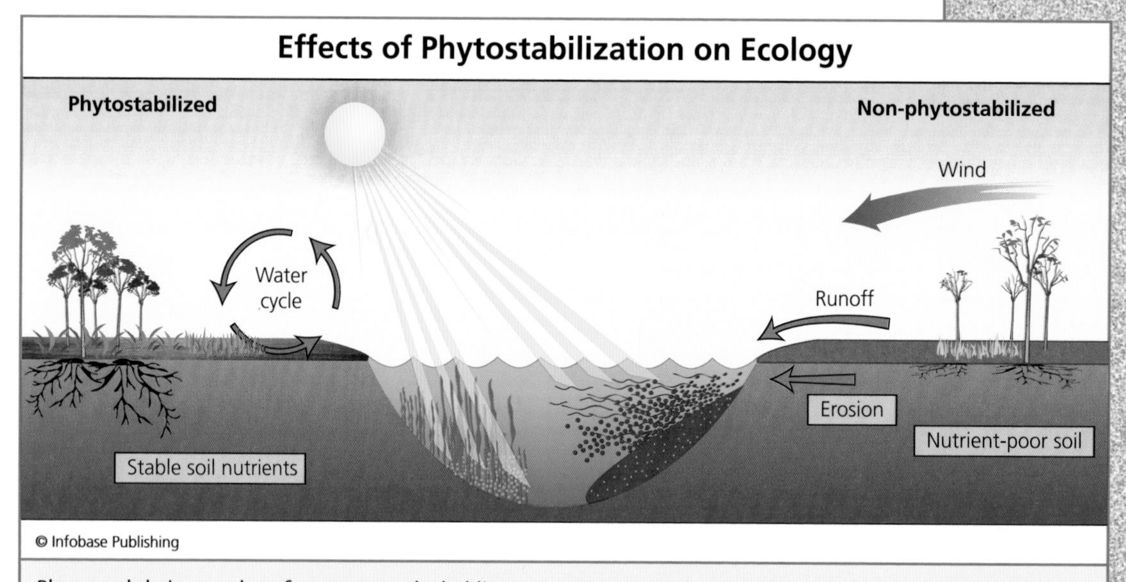

Plants and their roots benefit ecosystems by holding contaminants in place, enriching the soil, and decreasing erosion and runoff. Phytostabilization plants can be harvested after they have absorbed contaminants, and then the plants must be put in a landfill or incinerated.

The best vegetation for cleaning contaminated soil has the following three main attributes:

1. They are native to the local climate.
2. They are compatible with local soils.
3. They resist the area's pests and diseases.

As examples of these attributes, willow trees work best in temperate climates for stabilizing leachates from polluted areas, while citrus trees work better in warm to tropical climates and in dry soils to carry out similar remediation. Different types of plant life may be selected to provide specific advantages in phytoremediation. For instance, eucalyptus trees have deep root systems that grow through heavy clays and poor quality soils and may pull contaminants from groundwaters; poplars, cottonwoods, and aspens grow fast and adapt to a variety of climates; and root vegetables such as carrots, beets, and potatoes are easy to harvest.

In 2004 the Ford Motor Company employed a combination of New England aster, joe-pye weed, and leadplant to clean up polycyclic aromatic hydrocarbons (PAHs) that its steelworks had been releasing into the environment for years. These types of plant combinations may be more effective than plantings of a single species (monoculture) because the plant activities complement each other. Phytoremediation studies have found that a combination of two or more of the following plants provide the best activity for cleaning up metal-contaminated sites: Indian mustard, ferns, sunflowers, barley, hops, nettles, and dandelions.

For cleaning wetlands and ponds, plants that produce dangling root systems, such as sunflowers, are practical; the method is called *rhizofiltration*. Rhizofiltration has so far worked well in cleaning up organic chemicals and the radioactive chemicals strontium 90 and cesium 137, and treatment ponds and specialized greenhouses can be built for cleaning other specific contaminants from water. The treatment facility pumps a slow flow of contaminated water through troughs containing the dangling roots of plants suspended above. As the water flows, contaminants move from the water to the plant roots.

Studies on plants and trees in North America have shown that the best choices for phytoremediation have all or most of the following six characteristics:

- They grow fast and produce healthy plants.
- They consume large amounts of soil moisture and groundwater.
- They are perennial and adapted to winter conditions.
- They have deep and extensive root systems.
- They are able to grow in highly toxic places.
- They transpire water from the plant body over a long growing season.

The last of these points relates to an emerging technology called *phytovolatilization.* In this process the plant absorbs organic chemicals through its roots and transports them to the leaves. During their time inside the plant, contaminants break down into smaller molecules. The plant then releases these small detoxified compounds into the air as a vapor. Herbicides, organic solvents, and some munitions have been treated by this natural volatilization. Willows, poplars, aspen, cottonwood, grasses (fescue, sorghum, rye, Bermuda), and the legumes clover and alfalfa are good choices for phytovolatilization.

Another interesting variation on phytoremediation involves the use of fungi rather than leafy plants. This process, called *mycofiltration,* absorbs contaminants from soils or bodies of water. Many mushrooms and molds produce large underground mats of mycelia, which spread their thin filaments over an area that can cover many acres. These mycelial mats either spread over a body of water's surface or extend into soil to absorb chemicals, small particles, or toxic metals bound with dirt particles. The white rot fungus is an example of a mushroom currently used for degrading the long hydrocarbon chains of petroleum products and large molecule pesticides and herbicides. To do this, white rot uses part of the very same disease-causing enzyme system it uses to break hydrogen-carbon bonds in tree fibers. Once the infected tree has been robbed of its strong structural fibers, its wood contains only cellulose fibers, which gives the wood a white appearance.

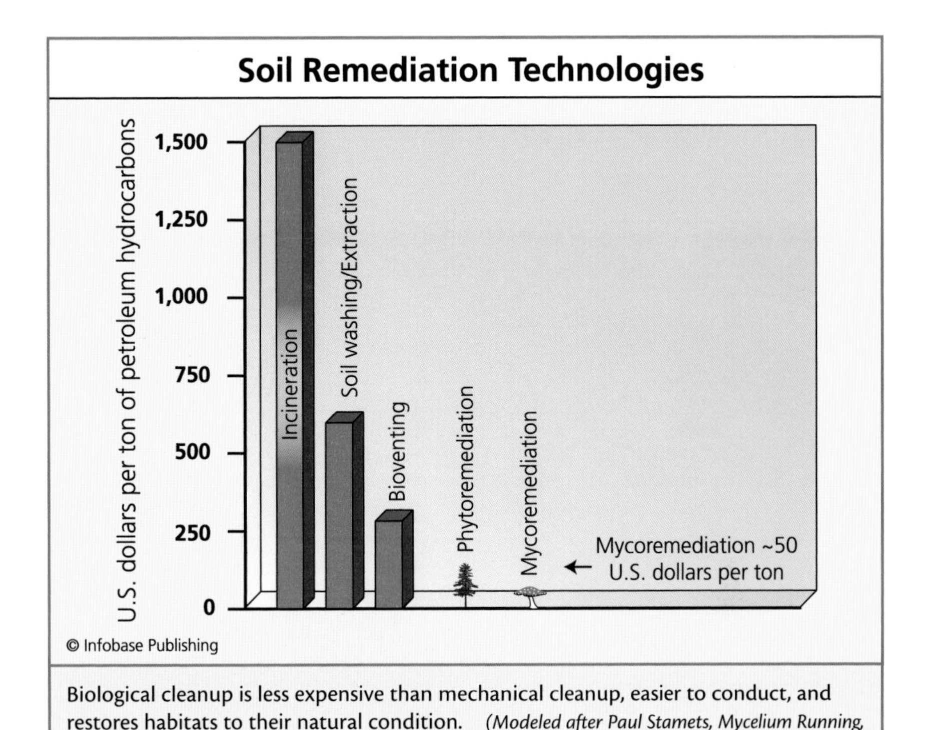

Biological cleanup is less expensive than mechanical cleanup, easier to conduct, and restores habitats to their natural condition. *(Modeled after Paul Stamets, Mycelium Running, Berkeley: Ten Speed Press, 2005)*

Remediation using fungi has several advantages. First, it is inexpensive and requires almost no maintenance once the fungi begin to grow, and they usually grow very quickly. Second, many fungal species act as hyperaccumulators for arsenic, cadmium, lead, mercury, copper, chromium-copper-arsenic complexes, and radioactive cesium. Third, types called saprophytic fungi readily degrade organic matter in the earth and grow throughout the topsoil. Saprophytic fungi therefore absorb shallow contamination from surface spills. Finally, fungi adapt to a wide range of environments from warm to below freezing. Fungal enzymes often work at low temperatures where plant and tree systems are inactive. In addition to metals, mycologists have discovered fungi suitable for extracting other contaminants from land, including benzopyrenes, anthracenes, chlorinated organic compounds, PCBs, PAHs, dioxin, organophosphates, and trinitrotoluene (TNT).

Phytoremediation plants and fungi receive a score based on their capacity to accumulate metals. The numeral score is called a bioaccumulation factor, which is the ratio of metal concentration in the plant to its concentration in soil. Such calculations can be made for vegetables and fruits sold in groceries, though this practice would receive considerable resistance

Fisher Creek in Montana's Gallatin National Forest contains acid drainage from an abandoned mine. Certain bacteria can be used to neutralize the acids and detoxify the toxic metals. *(Westerners for Responsible Mining)*

from growers and grocery retailers. In environmental medicine, bioaccumulation factors are calculated for concentrations of hazardous chemicals in human, animal, fish, and shellfish tissue.

CONCLUSION

Natural processes carried out by microbes, fungi, plants, and trees clean up contaminated sites and disrupt the environment less than excavation and other mechanical methods. Bioremediation is the use of microorganisms to clean up contamination; phytoremediation is the use of plants. These processes can detoxify, degrade, immobilize, or extract contaminants, depending on the nature of the pollution and the type of plant or microbe. Microbes work best in degrading or detoxifying organic compounds and detoxifying metals. Plants with extensive root systems work best in stabilizing or extracting contaminants. Certain hyperaccumulator plants possess superior capacity to absorb contaminants from the earth.

Phytoremediation contains several specialized categories. Phytostabilization is a method by which plants hold contaminants and do not allow the contaminants to move in the soil or water. In phytoextraction, plants pull the contaminant out of the soil using their root systems. Rhizofiltration is a method of pulling contaminants from slowly flowing waters using root systems suspended in the water. Phytovolatilization uses the ability of plants to pull compounds in through their roots, then transpire the volatilized form of the compound into the air. Finally, mycofiltration uses fungi rather than leafy plants to absorb contaminants from soil.

Microbes or plants may be genetically engineered to contain genes that improve their remediation activity. The first genetically engineered organism used on a large scale in agriculture was *Bacillus thuringiensis* bacteria. This species, nicknamed Bt, provided a gene that allowed plants to produce their own pesticide, thus reducing the need for chemical pesticides. Genetic engineering using bacteria and plants has the potential to improve both bioremediation and phytoremediation, respectively. Genetic engineering confronts the problem of public resistance, however, because people may fear the perceived or real threats that engineered species would have in the environment.

Certain extremophile bacteria and hyperaccumulator plants can carry out contamination cleanup without the need for bioengineering because they are already adapted to environments containing high levels of toxic materials. Research in both these areas holds promise for new natural cleanup methods.

Bioremediation and phytoremediation are gentle to ecosystems and preferred by most communities. These methods work slowly, however, and may not always be suitable for large cleanup tasks or cleanups that require fast action. Current contamination cleanup often uses a combination of natural systems and mechanical methods; biopiles and bioventing are two such natural-mechanical combinations.

Even with their advantages, natural cleanup systems do not return a site to its original pristine condition. For this reason bioremediation and phytoremediation may be best used to complement other mechanical approaches, such as excavation.

Oxidation Technology

Oxidation and *reduction* reactions are chemical processes in which compounds transfer electrons between them. Oxidation reactions play an important role in environmental science because oxidation often neutralizes chemicals that are difficult to excavate or treat by other means. In oxidation a compound or molecule loses one or more electrons and thus becomes oxidized; in reduction, electrons are accepted by a compound or molecule. The term reduction is used because the electron-accepting chemical becomes more negatively charged. Surprisingly, oxygen is not required for oxidation reactions, but it is a very common electron acceptor in chemical and biochemical reactions. An example of such a chemical oxidation is the rusting of a nail, wherein oxygen combines with iron to form iron (or ferric) oxide (Fe_2O_3).

In oxidation-reduction reactions, simply called redox reactions, one molecule is oxidized by giving up an electron to an electron acceptor and a different molecule is reduced by accepting an electron. This is illustrated by the redox reaction between one molecule of zinc (Zn) and one molecule of copper (Cu):

$$Cu^{2+} + Zn <--> Cu + Zn^{2+}$$

Here, zinc is oxidized when it loses two electrons to copper; copper is reduced by accepting the two electrons. The half-reactions of the total redox reaction are as follows:

$$Zn = Zn^{2+} + 2e^{-} \text{ and } Cu^{2+} + 2e^{-} = Cu$$

Oxidation of hazardous materials has currently become one of the principal methods of pollution cleanup, and it works on compounds far more complex than zinc, copper, or other toxic metals. Oxidation-reduction reactions also detoxify phenol compounds, organic solvents, volatile organic compounds, and cyanides.

Metals are very reactive in redox chemistry, so oxidation technology has worked well at sites contaminated with metals. Very large cleanup jobs containing metals mixed with other contaminants typically employ a combination of cleanup technologies to supplement chemical oxidation, such as physical excavation and biological remediation. Oxidation is attractive because it works at the contaminated site, meaning it is a type of in situ cleanup, and oxidation does not require heavy equipment to excavate materials or trucks to transport the materials to disposal sites. Oxidation is also a safe method for many cases in which the redox reactions produce water and harmless organic compounds.

Oxidation technology employs three different methods: thermal (by heating), chemical (also called *catalytic oxidation*), or biological. Thermal methods use high temperature to drive the redox reactions, while catalytic methods rely on strong chemicals. Thermal or catalytic oxidation is simple and inexpensive compared with other types of cleanup technologies. These reactions are also often fast enough to reduce the hazard within seconds.

The advantages of redox reactions have made oxidation technology one of the fastest-growing methods for sites needing immediate cleanup, as this chapter will show. The following discussion describes how oxidation chemistry developed into a method for cleaning up wastes and it describes different types of oxidation technology.

WASTE OXIDATION HISTORY AND CURRENT METHODS

Early pollution cleanups seldom converted contaminants to a less hazardous form. Untreated soil and sludge were excavated with backhoes, picked up, and hauled away, usually to be buried deep underground or dumped into landfills. This process could be very labor-intensive at large contamination sites, and as industries grew and wastes mounted, cleanup projects needed more compact systems.

The chemical industry matured as the industrial revolution blossomed. Before the 1970s chemical manufacturers enjoyed a favorable reputation in the public's eyes. New chemicals meant new conveniences. Pesticides, preservatives, and new plastics replaced or helped many manual tasks, and consumers began to rely on new chemical products to make life easier. The chemical industry's stature grew from the 1930s through the 1970s as it provided the public with convenient packaging, foods, drugs, and household items. In this period, Western society shifted from an agriculture-based economy and to an industrial economy, partly due to new chemicals. Incinerators worked overtime eliminating industrial wastes and by-products. What better way to reduce waste than to burn it until it disappeared in a puff of smoke? But in time that puff of smoke came to symbolize not efficient waste treatment but the health threats coming from toxic emissions.

Incineration, even with its pitfalls, served U.S. companies for decades. The fledgling nuclear industry joined the trend by using incinerators to destroy low-level radioactive substances. But the public became more and more opposed to the smokestacks, and by 1955 Congress passed the Air Pollution Control Act, which provided states with funding and technical assistance in air pollution prevention. The new law did not include air quality standards—that is, the maximum allowable levels of air pollutants—so a 1970 amendment to the act set forth specific limits for air pollutants that all states were to follow. Even with the government's belated attempt to clean up the air, the public had had enough of blackened skies. Burning wastes could not continue as the country's main waste-disposal method. Oxidation technology offered precise chemical reactions to detoxify hazardous wastes.

Today's chemical oxidation technology is divided into three groups according to their effectiveness in detoxifying compounds: primary degradation, defusing oxidation, and mineralization. Primary degradation involves a structural change in a compound, which makes it less toxic. In primary degradation toxicity is reduced but not necessarily eliminated. Defusing oxidation changes a hazardous compound so that it is completely detoxified, though there are a few instances in which toxicity is greatly reduced but not eliminated. In mineralization, an organic pollutant is degraded to carbon dioxide (CO_2) and all toxicity is gone. Chemists specializing in oxidization technology must take care because some oxidations cause a structural change in a compound that makes it *more* toxic than its original structure. Technicians in oxidation science must therefore have good chemical training before setting up a cleanup project.

Today's in situ chemical oxidation methods are primarily for treating persistent organic compounds in groundwater, soil, or air. Each operation

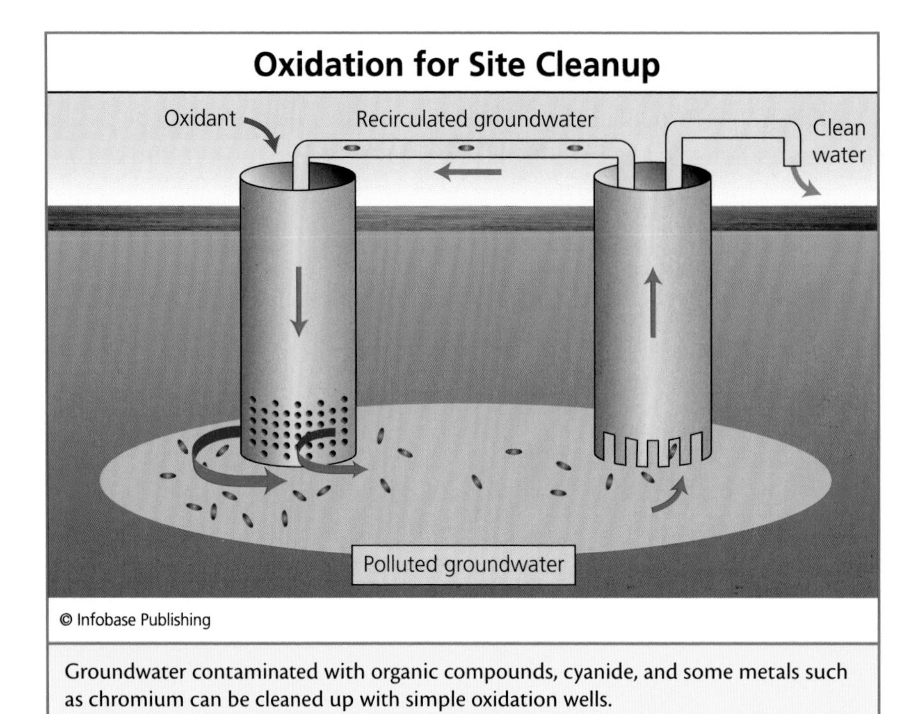

Groundwater contaminated with organic compounds, cyanide, and some metals such as chromium can be cleaned up with simple oxidation wells.

begins by drilling at least two wells into the contaminated soil or groundwater. Pumps then deliver an oxidizing agent, also called an oxidant, into the injection well. Underground the oxidant reacts with the hazardous chemicals and after a short period of time the liquid exits through the second well. Technicians often reinject the mixture for additional treatment until all the contaminants have been neutralized. Examples of contaminants that are commonly treated by oxidation are shown in the table on page 77.

Almost all in situ oxidations use oxidants to help the reactions taking place in soil or water. The most effective oxidants for contamination cleanup are the following: permanganate salts such as potassium or sodium permanganate; hydrogen peroxide; calcium peroxide; ozone; and sodium persulfate. Hydrogen peroxide (H_2O_2) depends on *Fenton reactions,* which are a series of linked reactions between the oxidant and an iron *catalyst*; a catalyst is a chemical—iron, manganese, and nickel are the most common—that speeds the reaction rate. Fenton reactions proceed quickly but they can be quite violent if not carefully controlled. At times the force of Fenton reactions drives matter up through the wells and spreads the pollution farther.

No cleanup technology is perfect, and oxidation has a number of factors that must be considered before beginning to use this technology. Permanganate compounds are persistent and can be toxic to native microorganisms, which are needed to help clean the site through bioremediation. Hydrogen peroxide and ozone have a contrasting problem; they dissipate quickly and so do not always reach the contaminants. In addition, each oxidant works better against certain types of compounds than others. For example, permanganates work poorly against benzene, PCBs, carbon tetrachloride, and dichloromethane, and hydrogen peroxide is ineffective against pesticides, yet ozone works well against most pesticides. Most oxidants are also corrosive chemicals that require careful handling and safety precautions. Choosing the correct catalyst is critical to ensure all of the hazardous compounds have been detoxified in the redox reactions because incomplete redox reactions, like incomplete incineration, create new toxic chemicals that can be dangerous at very low concentrations.

Major Contaminant Classes Treated with Oxidation

Soil and Groundwater Contaminants	
pesticides	methyl tertiary-butyl ether (MTBE)
petroleum hydrocarbons, greases	phenols
benzene, chlorobenzene	polychlorinated biphenyls (PCBs)
explosives	polycyclic aromatic hydrocarbons (PAHs)
Air Contaminants	
nitrogen oxides	volatile organic compounds (VOCs)
particulates, metals	diesel exhaust

Though chemical oxidation is the main means of using redox reactions for neutralizing toxic substances, microbes also biologically oxidize certain organic compounds. Microbes use oxygen for converting sulfur compounds to safer forms. When oxygen is not available for this conversion, anaerobic microbes, which do not require oxygen to live, take over and use nitrate as an electron acceptor rather than oxygen. Currently wastewater treatment plants are the only places where biological oxidation is used on a large scale. Wastewater treatment tanks containing bacteria detoxify a diverse mixture of compounds, including pesticides, organic solvents, and petroleum products. At hazardous waste sites, bacterial processes work slower than chemical oxidation. Conditions at a hazardous site can furthermore inhibit the natural enzymes so that some contaminants remain untouched. At this time, chemical oxidation is the better choice in contamination cleanup when compared with biological oxidation.

OXIDATION REACTIONS IN WASTE DETOXIFICATION

Many pollutants detoxified through redox reactions produce a molecule called a hydroxyl free radical. Free radicals are highly reactive chemicals that detoxify organic contaminants in a number of ways: by breaking an aromatic ring; by removing hydrogen; or by transferring electrons to create an unstable ionized form of the compound. Oftentimes these free radical reactions convert an organic compound to water, carbon dioxide, and a salt, in a process called *mineralization*. Mineralization is one of the most thorough and safe ways to eliminate contamination.

Though microbes do not work as quickly as chemicals, microbes detoxify contaminants in a safe manner because they mainly use mineralization. One example of a detoxification process carried out by bacteria takes place in areas contaminated by mine tailings. Mine tailings are toxic, acidic materials in the drainage from mining operations. Oxidizing bacteria help recover metals from tailings and desulfurize acids produced in bituminous-coal mining. (Bituminous coal is the most plentiful form of coal in the United States and is used for generating electricity.) In ore mining, *Thiobacillus ferooxidans* bacteria carry out redox reactions that

Ozone

Ozone is a gas made of three oxygen molecules. It occurs naturally in Earth's stratosphere—12 to 30 miles (20–50 km) above the planet's surface—or it is made by ozone-supply companies using chemical reactions. Ozone readily breaks apart into unstable free radicals consisting of hydrogen-oxygen plus an extra electron. Because the free radicals from ozone interfere with biological reactions, they are used to destroy organic compounds in drinking water and wastewater. Since its first use as a water purifier in France in the early 1900s, ozone is still the main treatment method for drinking water and wastewater in Europe.

The Ozone Cycle

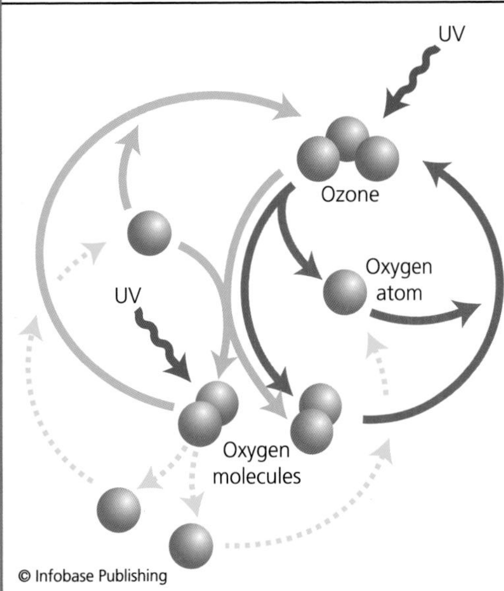

Ozone gas provides benefits and hazards. Ozone in the upper atmosphere protects the Earth by absorbing ultraviolet rays from the Sun, but in the lower atmosphere ozone forms hazardous smog. Ozone also is used as an effective chemical for disinfecting drinking water.

Ozone also protects life through its reactions in the atmosphere. Ozone forms when short-wavelength ultraviolet (UV) light irradiates oxygen molecules. (This is called a photochemical reaction because it involves energy from light.) The newly formed ozone absorbs long-wavelength UV light and thus prevents large amounts of UV from reaching the Earth's surface. Some UV light is essential for animal life—it enables the conversion of sterols to vitamin D—but excessive exposure to light in the form of UVB radiation leads to cell damage and skin cancer.

In the lower atmosphere, the troposphere, nitrogen oxides from burning fossil fuels react in sunlight to make ozone. This type of ozone presents a health hazard rather than a benefit because tropospheric ozone contributes to smog, which in turn aggravates respiratory ailments, including asthma. Ozone's diverse set of reactions provide a good example of the highly reactive character of certain chemical oxidations.

Decreasing Pollutant Toxicity

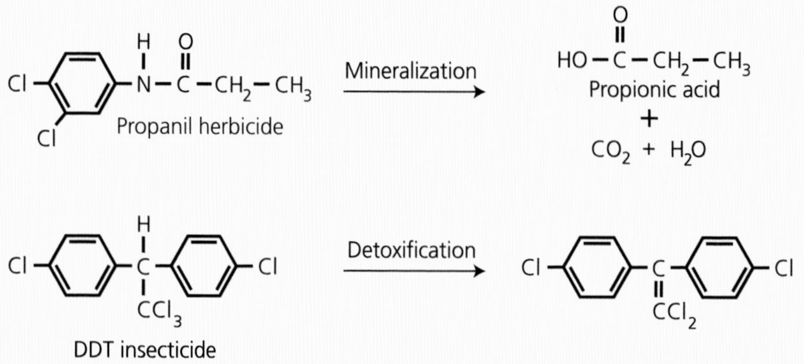

Propanil herbicide → Mineralization → Propionic acid + CO_2 + H_2O

DDT insecticide → Detoxification

In detoxification, bacteria and fungi in soil make chemicals less toxic by removing a part of the compound, rearranging it, or even adding a new molecule to it. In mineralization, microbes degrade chemicals completely to harmless compounds.

help recover the smallest amounts of copper, gold, and uranium remaining in the treated ore materials.

Acidophiles (bacteria that thrive on acids) such as *T. ferrooxidans* also produce a cleaner-burning form of coal by removing sulfur from high-sulfur coals. As a result these bacteria help reduce the amount of the greenhouse gas sulfur dioxide (SO_2) released into the atmosphere from burning coal. Oxidation plays a role in the cycling of another greenhouse gas, discussed in the sidebar "Ozone."

CATALYTIC OXIDATION

Catalytic oxidation is a process for making gas emissions less toxic. It depends on a catalyst, which enables the oxidation reactions to occur at 600°F–700°F (315°C–370°C), much lower than in thermal oxidation. Nickel, chromium, copper, rhodium, and manganese compounds are all fine catalysts, but platinum and palladium are the most efficient catalysts in use today; most high-efficiency industrial oxidizers and automobile catalytic converters use them. A converter's inner surface contains at least two of these metals; one serves as an oxidation catalyst and the other a reduction catalyst. When gas moves through the catalytic converter, oxygen and organic compounds react on the metal surface to form water vapor, carbon dioxide, and a small amount of hydrochloric gas, plus the release of energy as heat. This type of oxidation is currently the prevalent method for removing VOCs from industrial emissions and partially degraded fuel hydrocarbons and nitrogen oxides from car exhaust. Other compounds detoxified by catalytic oxidation are shown in the table on page 82.

Today's vehicles use catalytic oxidation in pollution control. Combustion engines normally produce nitrogen gas in addition to carbon dioxide and water, but inefficient engines do not burn fuels completely and emit a mixture of carbon monoxide, VOCs, and nitrogen oxides (NO and NO_2). Catalytic converters have played a critical role in lessening air pollution by taking these compounds out of car exhaust: The reduction catalyst

Mining creates various costs to the environment from habitat destruction to ecosystem pollution. In this picture a Superfund site in Crested Butte, Colorado, holds mining wastes called tailings. Poorly managed waste ponds leak hazardous chemicals into soil, groundwater, and runoff that carries the chemicals to surface waters. *(Coal Creek Watershed Coalition)*

POLLUTANTS TREATED BY CATALYTIC OXIDATION	
POLLUTANT CLASS	**EXAMPLES OF COMPOUNDS**
VOCs	trichloroethane, trochloroethylene, *BTEX*
semi-VOCs	PCBs, PAHs
extraction off-gases	various by-product gases such as VOCs
fuel hydrocarbon vapors	hydrogen-carbon petroleum compounds

reduces nitrogen oxides to harmless nitrogen gas (N_2) and the oxidation catalyst converts unburned hydrocarbons and carbon monoxide (CO) to carbon dioxide. Unfortunately, carbon dioxide is a greenhouse gas, so even so-called "clean car emissions" contribute to air pollution. Overall, however, catalytic converters have helped clean up some of the nation's worst air-polluted areas. One example of their use is given in "Case Study: Controlling Diesel Pollution in Massachusetts" on page 84.

THERMAL OXIDATION

Thermal oxidation depends on temperatures as high as 1,600°F (870°C) rather than on catalysts when destroying hazardous substances. The intense heat drives the oxidation of contaminants and vaporizes them. A collection unit then captures the vapors and cleans the emissions with a scrubber, which is a pollution-control chamber containing an absorbent material that retrieves specific gaseous compounds. After the underground heating and aboveground collection, the scrubber removes the evaporated contaminants and emits clean air. This method of thermal oxidation combined with the use of scrubbers has become an efficient method for removing VOCs from hazardous-waste sites.

A new type of thermal oxidation called flameless thermal oxidation destroys chlorinated organic compounds and aromatic hydrocarbons such as benzene and toluene. Flameless technology uses a ceramic bed heated with electric current up to 1,850°F (1,010°C) to vaporize contami-

An automobile catalytic converter. Simple devices such as catalytic converters on cars and trucks have helped decrease the volatile organic compounds present in vehicle emissions, but they also release carbon dioxide, a major greenhouse gas and one cause of global warming. *(L & G Auto)*

nants, which usually break down to carbon dioxide and water. Cleanup crews drill thermal wells into deep contamination sites and apply steam, hot water, or a vacuum to extract the reaction products. Newer methods involve using sonic waves to destroy contaminant compounds. A technology similar to thermal wells, called a thermal blanket, is used for surface contamination. Each method works in situ with virtually no dangerous end-products.

Cleanup teams using any of these oxidation methods strive to reduce the products of incomplete combustion (PICs). PICs result from reactions that stop before all of the contaminant has been destroyed. For instance, a faulty catalytic converter on a car can release as many pollutants as a car with no converter at all. Two major causes of incomplete combustion leading to PICs are insufficient reaction temperature and insufficient air supply. The Environmental Protection Agency (EPA) requires companies to monitor their emissions for PICs and correct any problem. Usually companies abide by this requirement by installing an afterburner on an

CASE STUDY: CONTROLLING DIESEL POLLUTION IN MASSACHUSETTS

In the 1990s, the Massachusetts Turnpike Authority (MTA) tackled the immense job of rebuilding Boston's major expressway. Before work even began on the Big Dig, as it came to be known, residents feared the environmental upheaval expected from the noisy, heavy equipment needed for excavation, demolition, construction, and street building. Air pollution would threaten a city already struggling with air quality standards. Early in the Big Dig, the MTA therefore unveiled a plan that would become one of the most innovative pollution-control programs in the history of the Clean Air Act.

The program required all diesel operators to follow certain guidelines for reducing noise, as well as diesel exhaust, airborne dusts, smoke, and odors. Many in Boston wondered if the plan could work, since it was based chiefly on behavior and common sense. Some of the actions truck and vehicle operators took were the following:

- turning off diesel engines when not in use
- turning off truck engines if idling in lines for more than five minutes
- using staging areas for trucks located outside residential areas
- locating equipment away from neighboring buildings' fresh-air intakes (for ventilation or air conditioning, or open windows)

To these actions the MTA added the most groundbreaking idea of all. In 1998 it began a voluntary program for diesel equipment and vehicle owners. All of the large off-road machinery was to be retrofitted with catalytic converters. Though catalytic converters had been developed in the 1970s at General Motors, they were mainly used on automobiles, not on big diesel equipment. The MTA encouraged owners retrofit their bulldozers, backhoes, excavators, cranes, and generators. According to the MTA, "Overall, the Big Dig retrofit program is now being used as

incinerator outflow pipe or on the thermal oxidizer. The afterburner then reheats the emissions to destroy any PICs.

INNOVATIONS IN POLLUTION CONTROL

Oxidation technology has led to a number of promising innovations for waste cleanup. Supercritical water oxidation (SCWO, also called hydro-

a model by regulatory agencies to encourage other construction projects to utilize retrofitted diesel equipment."

The Big Dig lasted until 2005 and overall the MTA's innovative program reduced daily carbon monoxide emissions by 93 pounds (42 kg), hydrocarbons by 30 pounds (14 kg), and particulate matter by almost 8 pounds (3.6 kg), and the converters ended up costing no more than 1 percent of the total cost of all the Big Dig machinery. Leaders in Boston took note of the Big Dig's success. In 2006 the city retrofitted its fleet of school buses. In an article published in the *Boston Globe,* environmental attorney Carrie Russell said, "This retrofitting is a great plan. But the reason retrofitting these buses is so important to children is that they are exposed, potentially for long periods of time, to bus exhaust. It will have an immediate improvement."

The tunnel and expressway construction project during the 1990s in Boston caused dirt, noise, and emissions pollution, but it also offered an opportunity for contractors to find innovations in pollution reduction. *(Built Environment)*

The Big Dig proved that people would be willing to volunteer to use environmentally sound measures. It gives hope that similar programs will be followed at all sites that depend on large and polluting machinery.

thermal oxidation) combines very high temperatures with high pressure. Pressures in the moisture-containing reaction chamber up to 360 tons per square foot (3,529 metric tons/m^2) accelerate the decomposition of hazardous compounds. SCWO is best suited for cleaning contaminated flowing waters of benzene ring compounds, dioxin, chloroform, and explosives and nerve gas. The U.S. Air Force, Army, and Navy have lately adapted SCWO methods to cleanup work on sites filled with explosives

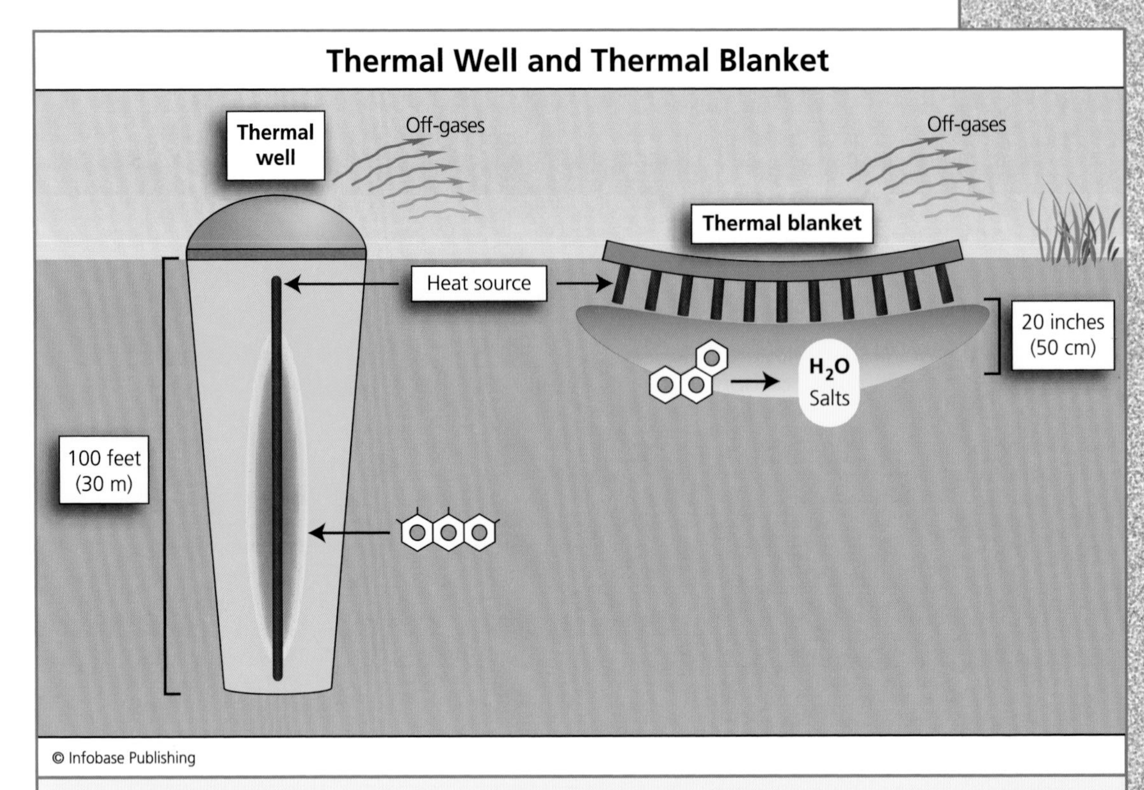

Thermal wells (for deep soil pollution) and thermal blankets (for shallow pollution) use desorption technology. They heat soil until contaminants vaporize, and then gas collectors or carbon filters capture the vapors. Wells and blankets clean up pollution directly in the soil without the need for excavation.

and munitions. Referring to these dangerous substances, the U.S. government's Pacific Northwest National Laboratory states on its Web site (URL: www.pnl.gov), "Supercritical water is now widely recognized for its capacity to destroy toxic or hazardous materials."

Other oxidation technologies on the horizon consist of electron beam, ultraviolet photo-oxidation, cerium-mediated electrochemical, and molten salt. Electron beam oxidation and ultraviolet photo-oxidation (UVPO) break chemical bonds by directing an electron beam at the chemical. The electron beam method works best on VOCs and VOC-like compounds. UVPO breaks carbon-hydrogen bonds in a process called UV photolysis, which is also valued in wastewater treatment for destroying the deoxyribonucleic acid (DNA) in microorganisms.

Cerium-mediated electrochemical oxidation can be a successful method in which the cerium acts as a catalyst in combination with an electrical current. This innovation reduces the heat required to run the oxidation reactions; high temperatures are very difficult to manage in in situ oxidation technologies. Heat-requiring systems have another obvious drawback: Heat production uses energy.

Molten salt technology also uses an electrical field for oxidative reactions. Molten salts such as aluminum oxide can attain very high temperatures, conduct electricity, and serve as a chemical catalyst. Already, molten salt technology has been useful for destroying chemical weapons. Molten salt technology is a safe method of cleaning up wastes, but this technology uses considerable energy to achieve the high temperatures needed for the melting step.

In all oxidation technologies, cleanup programs achieve the best results by taking into account the following six factors:

- type of cleanup task—solids, gases, or liquids
- depth of contamination in soils and sediments
- type of chemicals in the contamination
- selection of the proper oxidant for the type of chemicals to be treated
- properties of the oxidant in specific soil types, at various moisture contents and pH, and migration properties in soils and sediments
- selection of oxidation technology—thermal, catalytic, or new technologies

CONCLUSION

Chemical oxidation is an effective, fast, and safe method for cleaning polluted sites. It is used for shallow soils, deep soils, groundwaters, and sludges. Redox reactions destroy a variety of hazardous chemicals. The pollutant classes now treated by oxidation technology include chlorinated organic compounds, hydrocarbons, MTBE, explosives, and pesticides. Oxidants help drive the redox reactions forward, and often catalysts are included to accelerate the reactions and reduce the overall energy needs. In addition, oxidation is an in situ method, so has the advantage of cleaning up contaminants without the need for hazardous-waste transport and treatment at another site.

Oxidation converts hazardous chemicals to less toxic forms through chemical reactions involving electron transfer. Common end-products of oxidation are water, carbon dioxide, and a salt. Two main categories of oxidation carry out these reactions: thermal oxidation and catalytic oxidation. Thermal oxidation relies on high temperatures to initiate the process; catalytic oxidation requires less heat due to the presence of a catalyst.

Oxidation technology is the main method by which catalytic converters on cars reduce dangerous emissions in exhaust. Catalytic converters house the redox reactions taking place on a reactive metal surface, usually palladium or platinum, although several metals have been useful in catalyzing redox reactions. Whether oxidative converters are used in cars or on incinerators, the cleanliness of the emitted air depends on 100 percent combustion of the hazardous chemicals. Inefficient combustion and incomplete oxidation both contribute to the release of greenhouse gases into the atmosphere.

New technologies focus on alternative ways to supply energy to run the redox reactions. These include heat combined with high pressure, electrochemical energy, and energy from UV or electron radiations. Older and also emerging oxidation technologies play an important role in in situ pollution cleanup. In addition to oxidation's speed and economy, it is more specific to particular pollutants than general excavation or extraction methods. New methods may make oxidation technology even more efficient and less energy-dependent in the near future.

Brownfield Sites

Brownfield sites, or brownfields, are hazardous sites that have reusable value and are administered by the U.S. Environmental Protection Agency (EPA). Canada, Germany, and the Netherlands also have brownfields, and the United Kingdom operates a hazardous waste program similar to the EPA's brownfield program.

The idea of restoring a hazardous waste site to a safe place for people and animals took years to develop. In the 1960s environmentalists used a number of serious toxic waste mishaps to illustrate to the public the devastation hazardous chemicals caused to land, sea, and biota. For many years hazardous waste had been disposed of far from city limits, out of sight. In the 1950s and 1960s, urban populations spread into suburban areas and then rural land. Places once reserved for heavy industry and waste disposal suddenly became the backyards of new communities. Residents saw firsthand the particles coming from smokestacks or blowing off factory lots. They drove past landfills, decrepit lots, and incinerators every day as they went to work and school. The water tasted bad and often made people sick. The time had come for new and strong laws against environmental polluters.

Stricter air and water regulations enacted in the 1970s had an unexpected result. Some manufacturers were unable to pay for cleanup costs or the installation of cleaner equipment. Others were simply unwilling to meet the tougher laws. Manufacturing plants began shutting down. Over the next several years military bases also began closing for political reasons, leaving behind land polluted with explosives and munitions. Neighborhoods near abandoned industrial or military sites may have looked cleaner, but more sensitive analytical methods had become available to environmental science. With these improved methods and equipment, scientists found dangers hidden in the soils and groundwaters, in lakes and streams, and on breezes flowing through homes. Removing the sources of pollution had not removed the pollution itself.

Urban tentacles continued reaching farther outward in the 1980s and the value of land in the United States increased in the outlying areas of cities. Old factory sites and military bases became attractive to developers for housing and businesses. Many communities wanted more land for recreation, city parks, and golf courses. Homebuilders also desired homes along coasts, rivers, and lakes, places once occupied by factories or gunnery ranges. Those abandoned properties had turned into valuable pieces of real estate.

Restoration of polluted, abandoned land is the principle behind the EPA's brownfield program. The term *brownfield* was born in the late 1990s to describe industrial property not being used to its fullest potential because of contamination. (The term is used for describing contaminated land; the term *greenfield* describes uncontaminated or undeveloped rural lands.) The EPA designates sites as brownfields based on two criteria. First, there must be contamination on the site or merely the perception that contamination is present. For example, an abandoned metal-plating shop might be designated as a brownfield even without proof of the presence of heavy metals. Since scientists already know metal plating produces toxic waste, the empty building is perceived as contaminated and the local government may identify it as a brownfield. Second, brownfield law takes into account the potential economic value of the land if it were cleaned up and reused. Brownfield law is therefore a rare environmental law that considers economic factors in addition to environmental factors.

When land sits idle, the local government cannot collect taxes on it. Unused property becomes an economic burden for a community because it does not earn any revenue. For this reason many cities desire to claim brownfield land and restore it for city use.

Brownfields occur in rural, suburban, or urban areas, but the majority of brownfields in North America occupy land near urban and industrial centers. They are sometimes difficult to define for three main reasons: (1) diverse histories from one site to another and varying degrees of contamination on each; (2) information on the types and amounts of hazardous waste may be closely guarded by the previous owner; or (3) a site's history is simply not known. Companies, governments, nonprofit organizations, universities, or private investors all have the right to purchase brownfield sites, but information about privately owned brownfields is usually not available to the public. Therefore, people can only make an estimate of the number of sites currently in redevelopment under the brownfield program. The federal Government Accountability Office has estimated there are between 425,000 and 600,000 U.S. brownfields in existence today; the EPA puts the number at 500,000 to 1 million, if small, private sites are included. Lawyer Robert D. Fox, writing for the legal newsletter *Metropolitan Corporate Council,* described the number of brownfields this way: "Virtually every community across America, from the inner-city to suburban and even rural locations, contains abandoned or under-used properties stigmatized with real or perceived environmental contamination."

The EPA's brownfield program helps states, communities, or private investors plan cleanup activities. The agency helps devise a cleanup plan and may suggest ways to contain costs, but the government agency does not do the actual cleanup work. The new brownfield owner takes responsibility for carrying out the steps from assessment to reuse. Brownfield site owners also have access to grants from the EPA that provide money for hiring environmental experts, job training, and for investigating the newest cleanup methods.

The EPA's brownfield program differs from its Superfund program in at least one important way: Brownfield redevelopment is voluntary for the purpose of earning money, while Superfund is not voluntary. This chapter looks at the current state of the EPA's brownfield program and the steps that waste managers and the EPA use to plan a site cleanup.

THE HISTORY OF TOXIC SITE DESIGNATION

In March 1979 the Three Mile Island nuclear power plant next to the Susquehanna River in central Pennsylvania experienced a mechanical accident. A combination of equipment failures and worker error caused thousands of gallons of the plant's cooling water to pour from a faulty valve and the reactor core overheated, causing a partial meltdown of the core. Radiation escaped into the surrounding community of at least 2 million people with untold health effects. The reactor's workers stopped the leak and avoided a potential explosion, but two years after the accident most of the contaminated water still filled the containment building. An anonymous company spokesman said to the *New York Times*, "(the company) has no specific knowledge on what damage has been done to the inside of the reactor vessel." Not until 1993 did nuclear workers complete a $1 billion cleanup of fuel and 2.2 million gallons (8.4 million l) of contaminated water.

Soon after the events at Three Mile Island, another environmental calamity unfolded in a residential area near Niagara Falls in New York. The area, called Love Canal, had been built on land used as a waste dump by a local chemical company from 1942 to 1953. The city of Niagara Falls won the legal right to take over Love Canal but the community had little knowledge of the extent of the site's pollution. With limited environmental technology or know-how, city leaders supervised a sloppy cleanup that released tons of hazardous chemicals into Love Canal's surroundings. New York State Health Commissioner Robert Whalen announced in 1978 that the area had become "a great and imminent peril to the health of the public." In that year news organizations reported on the tragedy unfolding in upstate New York, and within the next two years, the state and the federal government evacuated the entire population of Love Canal. In the following decades to the present day, adults and children have fallen ill with unusually high rates of cancers, birth defects, and autoimmune disorders. The EPA eventually called Love Canal "one of the most appalling environmental tragedies in American history."

Three Mile Island and Love Canal highlighted as never before the grievous environmental harm, both potential and real, of uncontrolled contamination. The public's concern over environmental poisons had been growing before these events, but it renewed its urgency for stronger environmental enforcement. Most Americans pointed to industrial polluters as the greatest source of hazardous wastes, and by 1980 they wanted strict legislation against polluters. A *Time* magazine cover story that year cited an *ABC News*–Harris poll that reported that "despite a growing antagonism toward government regulation, 93.6 percent [of U.S. citizens] wanted 'federal standards prohibiting such dumping made much more strict than they are now'." In that year Congress responded by passing the Comprehensive Environmental Response, Compensation, and Liability Act, to be known as Superfund.

Love Canal became America's first government-designated environmental disaster and its earliest Superfund site. Unlike many hazardous sites that languish through years of little or no cleanup, an interesting occurrence took place at Love Canal: its eight-year cleanup project made the area safe enough to use again. Hundreds of its houses were decontaminated and new homes and offices were built. When families began returning to Love Canal, it had become clear that severely contaminated land could be cleaned up, restored, and reused. The story of Love Canal's surprising rebirth is detailed in the "The History of Love Canal" sidebar on page 94. Perhaps contaminated and abandoned properties and buildings had economic value after all.

With Love Canal's experience in mind, the EPA in 1995 introduced the Brownfields Economic Redevelopment Initiative. Hundreds of towns and cities soon identified sites within city limits as potential restoration projects. But how were lands to be designated as brownfields? The variety of pollutants and the degree of infiltration into land and water probably seemed boundless in those early days.

The term *brownfield* entered conversation for the first time in 1993 at the annual Conference of Mayors. Chicago's Mayor Richard M. Daley proclaimed his city alone had at least 2,000 sites that would meet the definition of a brownfield. Daley and other mayors visited the EPA in Washington, D.C., to discuss funding for brownfield cleanup and redevelopment. They also tackled the question of who would be held liable for the cost of each cleanup. There was, and still is, concern from developers who begin building on "clean" properties, only to discover large amounts of hazardous substances. The team of mayors and the EPA knew that innocent developers could hardly be held liable for another company's pollution. After nine years of lobbying for funding and a clear plan for liability, President George W. Bush signed a related act into law in 2002: the Small Business Liability Relief and Brownfields Revitalization Act. "The passage of this bill will help revitalize many contaminated sites and surrounding communities, generating sorely needed jobs and tax revenue and improving the environment," New Orleans Mayor Marc H. Morial said. "Mayors have been on the forefront of this issue, drawing national attention to the pervasive problem of brownfields and seeking creative ways to reuse these sites and make them more productive for their communities." In his keynote address in 2001 to the Urban Parks Institute, Daley gave a far more promising report than he did in 1993. "The City and the State of Illinois intend to acquire and clean up 2,600 acres. The first parcel, 117 acres, is being donated by the Belt

(continues on page 96)

THE HISTORY OF LOVE CANAL

In 1896 William Love began excavating a two-mile (3.2 km) connector between New York's upper and lower Niagara River. Love had been inspired by Thomas Edison's experiments with electricity, so the entrepreneur decided to build a hydroelectric plant on the canal to serve the nearby Niagara Falls area. At the start of his project, electricity was harnessed only as direct current and traveled distances of no more than a few miles. This limitation convinced Love that he would be in the envious role of sole supplier of electricity to Niagara Falls. While Love counted his future millions, 400 miles (644 km) to the south engineer Nikola Tesla debated with Edison the merits of alternating current for transmitting electricity. Edison soon agreed with Tesla's belief that alternating current provided a more efficient form of conduction. As each new electric company started business, it adopted alternating current as its standard. William Love lacked the finances to keep up with the latest in electrical technology, and by 1900 he had abandoned his unfinished "Love's Canal."

Hooker Chemicals and Plastics Company took over the canal in 1942 and quickly filled open trenches with chemical wastes. Until 1953 the pits brimmed with toxic sludges; then Hooker turned ownership of the canal over to the town's Board of Education for the price of one dollar. The chemical company also extracted a promise from the local government that indicated Hooker Chemicals knew more about the land's hazards than it disclosed: No lawsuits could be filed over any illnesses due to the company's wastes. City workers set about filling in the canal and built a school. Classes were held for the next 23 years in the neighborhood known as Love Canal.

New roads and building construction began in the early 1970s as the area's population grew. Ditches and wells were dug. Ground was excavated for foundations. Before long, chemical-laced waters hidden for so long underground began bubbling into lawns, gardens, basements, and the school playground.

In 1978 Lois Gibbs wondered why her children and others suffered an unprecedented string of illnesses. When she read a news article about the 20,000 tons (18,144 metric tons) of chemicals under Love Canal's homes, Gibbs made the connection between the hazardous wastes and her community's health. Though she had never before played the role of activist, she organized a homeowners association after the school board and city government failed to acknowledge a potential health disaster. In 1979 Gibbs stated to a subcommittee hearing in Congress, "I became involved in this situation after discovering that toxic chemicals were buried two blocks from my home and that these chemicals could be aggravating my children's health problems, one of whom attended the 99th Street School located in the

center of the dump." For two years Gibbs and other residents pleaded with town leaders to move the school and vacate the area. Through it all Gibbs endured death threats and physical and verbal attacks by those not willing to accept the truth about their home. Said Gibbs, "Probably the most difficult obstacle to relieving the problems at Love Canal has been 'being the first.' Neither the state nor the federal agencies who could help were responsible for the situation."

The rash of sicknesses in Love Canal spurred President Jimmy Carter to declare it a disaster. The Federal Emergency Management Agency (FEMA) stated in its 1980 assessment, "Love Canal Relocation Task Force Status Report," "On August 7, 1978, the President declared that the adverse impact of chemical wastes lying exposed on the surface, and associated chemical vapors emanating from the Love Canal Chemical Waste Landfill in the City of Niagara Falls, New York, was of sufficient severity and magnitude to warrant declaration of an emergency under the Disaster Relief Act of 1974."

The government bought 800 homes and, for the first time in the nation's history, it relocated 900 families due to a toxic waste disaster. The national attention on Love Canal probably provided the main incentive for a vigorous cleanup. Crews excavated and removed the solid wastes; they secured the dumping ground and covered it up. Workers also built drainage lines to carry liquid wastes to a local treatment plant. After ten years of cleaning and restoration at Love Canal, the state began selling newly built homes and decontaminated homes. The neighborhood became Black Creek Village, and Lois Gibbs went on to found the Center for Health, Environment and Justice.

Love Canal's restoration may be an ongoing experiment. The long-term health of the site's original families is unknown. Bitterness over the behavior of industries and the inertia of government remains, as it did decades ago. Niagara Falls resident Agnes Jones wrote to the *Niagara Gazette* in 1980, "There are a large number of forgotten families in the Love Canal story. Their health problems have not been addressed at all . . . A number of these families participated in the state Health Department blood testing program, and never received results. The health problems go to the complete range from cancer to skin rashes. These people are confused, hurt and frustrated." The disaster did, however, establish the fact that wastes move through the environment and dumping them in one place does not always assure the safety of the surrounding neighborhood. Love Canal's experience also confirmed that environmentally ravaged land and water can be cleaned and restored.

(continued from page 93)
Railway of Chicago as part of a mitigation requirement for filling wetlands in its suburban rail yard. Not far from there, the Ford Motor Company is building a new factory on a large brownfield site between two inland lakes. It will be surrounded with native plants, and an existing ditch will be turned into a meandering stream. Whenever possible, we try to create parkland on the site of new developments."

TYPES OF BROWNFIELDS

In 1999 the National Brownfield Association was created in Chicago to help mediate the planning by property owners, cleanup experts, and government in brownfield cleanup. Several bills have since strengthened brownfield law in the areas of funding and tax incentives. In 2002 the Small Business Liability Relief and Brownfields Revitalization Act was signed into law. By this law, developers and landowners are not responsible for any contamination discovered after they purchase property.

Brownfield projects today begin with risk assessment in which experts review the financial risks of the cleanup as well as the environmental risks of the pollution. High-risk sites are those that are heavily contaminated and pose an immediate health threat to humans and the environment. The EPA may change the designation of high-risk sites from brownfields to

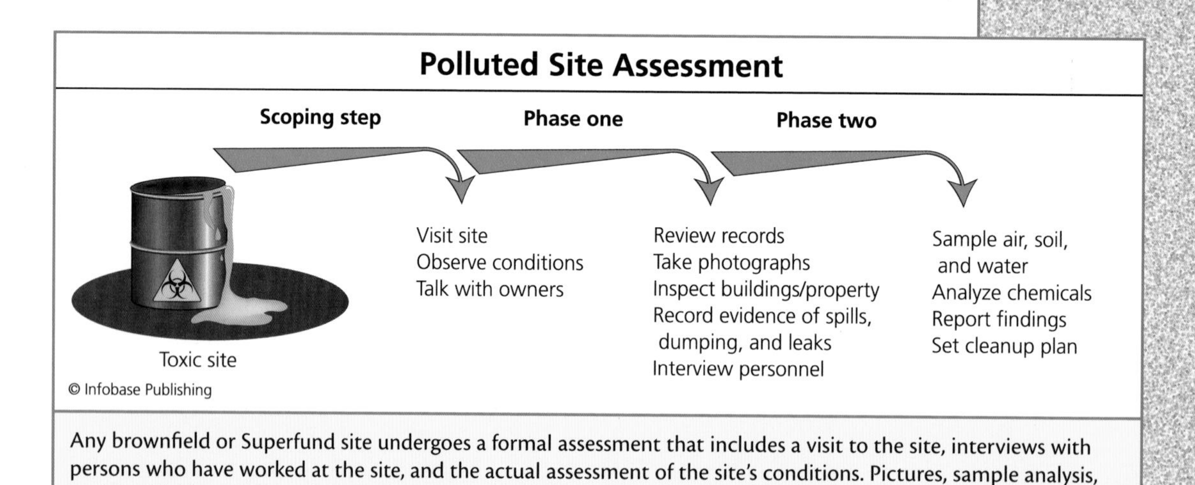

Any brownfield or Superfund site undergoes a formal assessment that includes a visit to the site, interviews with persons who have worked at the site, and the actual assessment of the site's conditions. Pictures, sample analysis, and detailed inspections of the buildings and company records are important in the overall assessment. All these steps occur before a cleanup plan is written.

Superfund sites. Properties that remain identified as brownfields undergo a cleanup plan, but only if the owner feels the costs of the cleanup are manageable. Most of the current brownfield sites in the United States are

The Pullman factory complex in 1920. From the 1880s to 1982, the Pullman site in Chicago produced railroad cars. The site also received sewage waste and liquid waste from a paint factory. The site now contains excessive levels of 23 different hazardous chemicals from previous dumping or industrial operations, including organic solvents, pesticides, and toxic metals. In August 1997 the U.S. Environmental Protection Agency designated the Pullman complex a Superfund site. *(Pullman Virtual Museum)*

designated as low-risk. This means they hold moderate or small amounts of contamination within a confined area, and these characteristics make them uncomplicated to remediate.

An environmental assessment takes place at the same time as the financial assessment. An environmental specialist samples the land, soil, and air to determine the amount of pollution present. This entire process of visiting the site and taking samples is called scoping. The person conducting the assessment then writes an environmental assessment report, which summarizes all the findings including the analytical test results. The report describes the potential dangers of the contamination and it

USED MOTOR OIL

People improperly dispose of 180 million gallons (681 million l) of used motor oil in the United States each year, mainly after changing the oil and oil filter in their own vehicles. An unknown additional amount enters the environment through accidental spills on private property and at gas stations and auto repair shops. One gallon (3.8 l) of used motor oil can contaminate 1 million gallons (3.78 million l) of freshwater, so it may not be surprising that it causes almost 40 percent of the pollution in North America's lakes, rivers, and wetlands.

Most motor oil pollution comes not from gigantic oil tanker accidents but from rinsing oil off paved surfaces into storm drains. From there, it flows to bodies of water and creates a sheen surface that blocks sunlight needed by aquatic plants. As the plants die, oxygen depletes and fish and shellfish suffocate. Oil-coated feathers cause shorebirds and ducks to lose body heat, as well as buoyancy in swimming and aerodynamics during flight. Furbearing animals that depend on lakes and rivers for diet or habitat—beavers, otters, muskrat—also lose heat when oil congeals fur. On land, oil interferes with nutrient and water uptake by trees and plants in *riparian habitats.* Chemically, oil adds toxic cadmium, chromium, lead, arsenic, zinc, benzene, and dioxins to ecosystems.

The EPA offers information about motor oil pollution to the public through the "You dump it, you drink it" campaign, which explains the correct steps for disposing and recycling used motor oil. EPA rules state that all auto service centers must keep used oils from mixing with other chemicals such as antifreeze, paints, and fuel. Waste oil must be stored in secure containers and shipped to a used-oil refinery. Car owners should not dump oil into drains, but rather take it and used oil filters to a gas station or service center. Some used-oil refineries accept waste oil directly from car contains recommendations for cleanup. Every brownfield landowner uses the environmental report to develop a cleanup plan that will follow all local and federal environmental laws. In the United States and Canada, government oversees almost every step of planning to help owners and developers finish the job properly. In the United States the EPA offers advice on cleanup based on past brownfield projects. The advice includes information on different types of chemicals likely to contaminate brownfields. Some of these contaminants are listed in the table on page 100.

Despite public perception, not all brownfields come from large industry. Some small businesses also produce dangerous chemicals.

owners. People today have little excuse for pleading ignorance about proper handling of waste oil; every state and most towns offer information on motor oil disposal and recycling. The organization known as Earth911 (URL: earth911.com) and the American Petroleum Institute (URL: www.api.org) also provide advice on motor oil recycling.

Used motor oil is re-refined at specialized refineries that clean it and replace additives to make a product identical to the lubricants made from crude oil. New oil from crude oil costs 15 to 50 percent more to produce than re-refined oil, and overall the recycled oil saves the automotive industry billions of dollars annually. Recycling plants first dewater dirty oil by holding it undisturbed in large tanks until the free or unbound water separates and falls to the bottom of the tank. This water is then collected and sent to a wastewater treatment plant. The next step involves heating the recovered oil and treating it with strong acids and detergent that cause all oil-bound water to form an aqueous layer. This water settles to the bottom, taking with it the acids plus metals and other contaminants, which are then removed, dried, and incinerated. The partially clean oil enters a de-asphalting step in which it reacts with propane gas. A clean oil phase and a dirty asphalt phase form. Finally, refiners distill the clean oil to make various oil grades, such as heavy, medium, or light, for different types of engines.

Re-refining can be thought of as a way to "close the oil loop." It lessens the world's demand for crude oil and enhances the proper care of used motor oil by businesses and individuals. Re-refining uses oil in a sustainable fashion, creates energy from waste, and conserves the world's oil reserves. But it is a one-way process; once recycled oil has been burned, refineries cannot process it further and the used oil must be incinerated.

TYPES OF BROWNFIELD CONTAMINANTS	
POTENTIAL CONTAMINANT	EXAMPLES OF SOURCE FACILITIES
fuel	service stations, airfields, bus facilities, heavy machinery storage yards, car lots, parking lots, truck stops, railroad yards, oil production facilities
chemicals or solvents	factories, military bases, dry cleaning facilities, pulp mills, chemical warehouses, abestos-contaminated buildings, salvage businesses, paint shops
heavy metals	ore mining sites, quarries, metal plating businesses, electronics and component manufacturers, automobile salvage, metal recyclers, wood preservers

Without proper handling and disposal of wastes, enterprises such as nail salons, copy shops, tire shops, cement plants, machine shops, and auto repair garages may leave behind hazardous materials when they move or go out of business. As the "Used Motor Oil" sidebar illustrates, even small drops of materials can build up over time into significant problems for ecosystems. Research laboratories, university laboratories, and hospitals are also potential brownfields. The EPA categorizes brownfields by industry type, as shown in the following table. Classifying brownfields by industry speeds the planning step because it tells the landowner the type of contamination to be expected on the site. Knowing the type of contaminants in turn helps determine the best cleanup technologies.

The brownfield's location often comes into play when developers, government leaders, residents, and environmentalists hold different opinions on the best use for restored land. The real estate industry and local government might stress the potential profits to be accrued by developing the property. For example, a brownfield in a depressed inner city might receive new offices and shops and bring other benefits to the community as part of a revitalization agenda. Outside the cities, many parks, preserves, wil-

EPA Brownfield Categories	
Type of Brownfield	**Cleanup Tasks**
community	small-scale projects needing excavation and removal or on-site containment
mill sites	removal of asbestos, VOCs, PCBs, metals (textile mills); wood treating chemicals, VOCs, fuels, metals, creosote, dioxins (wood and paper mills); petroleum-based products, PCBs, slag, underground storage tanks (iron and steel mills); Numerous buildings on large sites
mining sites	acid tailings, PCBs, heavy metals in remote sites; surface and groundwater contamination
portfields (ports and harbors)	oils, ballast, ship scrapings, paints, solvents in large underwater areas
petroleum sites	refineries, storage tanks, automotive plants, service stations; numerous structures on large areas connected with tanker shipping ports and railways
railfields	petroleum products, herbicides, pesticides, metals, creosote, industrial cargo on large areas covered with rails and buildings; right-of-way stipulations
underground storage tank (UST) sites	buried storage tanks with mixed contents, contaminated soils, sediments, and groundwaters

derness areas, and open spaces are best left undeveloped as greenfields in order to protect native ecosystems. In 2007 Rosanne Sanchez, manager for the Phoenix, Arizona, brownfields program, explained to the *Brownfield News,* "You have to be very upfront about what you intend to do, and bringing the community in during the early stages of a plan is extremely important."

The Hanford Superfund site cleanup along the Columbia River in Washington. Some Superfund sites and brownfields exist in places that would otherwise be desirable locations for homes and businesses. *(EPA)*

CLEANUP AND RESTORATION

After the risk assessments have been completed, the cleanup manager writes a restoration plan. The cleanup team consults the U.S. Geological Survey databases on the land's topography, erosion patterns, sediment movements, and surface and underground water flow, and then summarizes these in the restoration plan. The plan also outlines the safety measures to be taken, usually starting with a fence around the entire cleanup site.

Next, the restoration plan identifies the technologies best suited for the task. Cleanup experts decide whether the cleanup will be in situ or ex situ, meaning at another location. Often cleanup is a mixture of both. In situ methods reduce chances for spills and may save on costs. Ex situ methods, however, allow work to continue while excavated materials are treated elsewhere, speeding the overall cleanup.

Workers follow the steps described in the restoration plan once the project begins. They first separate contaminants by type, if possible: metals, liquid organic compounds, solid chemicals, hydrocarbon fuels, pesticides, explosives, and strong acids and bases. Compatible chemicals are consolidated before being treated or shipped. Sometimes a low-risk brownfield is restored to a safe condition merely by putting the hazardous materials into storage containers and removing them. A procedure called *capping* contains larger areas of surface contamination by using an impermeable liner to cover the contaminated area and prevent erosion. Underground barriers confine subsurface contamination and prevent migration or leaching. For example, a grout curtain is a barrier injected into bedrock surrounding a contaminated area. After hardening, it forms a wall that keeps the hazardous materials in place. Steel or iron sheets have also been used as barriers; slurry walls made of bentonite clay stop the movement of contaminated groundwaters in a similar fashion.

Drums or barriers cannot contain the largest brownfields. These sites require advanced technologies: excavation with soil washing or thermal desorption; incineration; chemical or biological extraction; soil vapor extraction; oxidation/reduction; precipitation, adsorption, or other physical separation techniques; in situ thermal methods such as *pyrolysis;* and chemical or biological stabilization. A broad choice of technologies is also available for liquid contaminants: *air sparging,* solidification, injection wells, and oxidation/reduction or ultraviolet oxidation.

The EPA encourages the use of bioremediation and phytoremediation whenever possible. Many residents prefer vegetation to detoxify contaminated brownfields rather than contend with the sight and noise of lengthy mechanical cleanups. Other communities by contrast may not be willing to give biological cleanup the time it needs to detoxify hazardous chemicals. These communities restore their brownfields by the "scrape and pave" approach. Backhoes and bulldozers remove the mess (scrape) and then construction crews build new parking lots, roads, and buildings (pave). Despite the occasional need for heavy-handed cleanups such as scrape-and-pave, an increasing number of communities today prefer green technologies for all or part of their brownfield project.

Abandoned buildings present workers with a unique cleanup task. Contaminated buildings undergo assessment the way contaminated land

Wetlands have the ability to clean natural waters and soils of pollutants. These wetlands in Manitoba, Canada, contain slow-flowing water, plenty of natural vegetation, and a continuous ecosystem with little human influence: the perfect conditions for pollution cleanup. *(Manitoba Liberal Party)*

and water do, but the methods for containing the hazards may be different. Brownfield crews prevent hazardous substances from leaking from buildings in a variety of ways, including:

- containment of all dusts with impermeable liners
- collection in leakproof containers
- repair of the building's structural damage
- use of a tracking system to prevent accidental losses of chemicals

Brownfield projects last from a few months to several years. After the cleanup is finished, an environmental scientist again assesses the soil, water, and air. The owner can begin redeveloping the land only if the tests show that the site no longer contains dangerous levels of hazardous chemicals. Redevelopment is important because it helps the owner and the project's investors recoup their investment. To speed the time between the cleanup's start and the point at which the property makes money, brownfield developers increasingly coordinate cleanup with new construction. In other words, the developer begins building on the parts of the property that have been cleaned even while other parts of the site still contain contaminants. This approach is called *integrated cleanup-redevelopment.* At present local governments and brownfield developers believe integrated cleanup-redevelopment is the only cost-effective way to restore most brownfields.

The EPA deems a brownfield cleanup complete when soil, water, and air testing results meet local, state, or federal limits. A state or local government office then issues a letter to the developer that declares the brownfield project a success. This letter, called a No Further Action letter, releases the developer from further cleanup responsibility. If a community tries to hold a brownfield developer responsible for any future problems at the site, the No Further Action letter protects the developer from that risk. Little wonder that it seems that once brownfield redevelopment projects have started, liability claims follow quickly before the EPA issues the No Further Action letter. As "Case Study: Responding to MTBE Pollution" on page 106 shows, hidden hazards have a way of appearing many years after contamination occurs.

Brownfield cleanups involve many aspects of environmental law. Developers rely on legal experts to guide them through all the legal requirements they must meet throughout the cleanup. Private legal firms specializing in environmental law, and the Environmental Law Institute (URL: www.eli.org) provide legal assistance in the world of brownfields. In addition the Institute of Professional Environmental Practice (URL: www.ipep.org), a part of Duquesne University in Pittsburgh, employs a staff of certified professionals in environmental policy who serve as a resource for brownfield developers.

REAL ESTATE AND THE ECONOMY

Students of environmental science would be foolish to expect all economic decisions to take a backseat to the world's ecology. Commerce has contributed in many ways to environmental crises, but there is no returning

(continues on page 108)

CASE STUDY: RESPONDING TO MTBE POLLUTION

In the 1970s the nation's oil refining industry responded to new air quality requirements put forth by the Clean Air Act by adding a small water-soluble hydrocarbon to gasoline to replace lead, which was incompatible with the catalytic converters being installed in new vehicles at that time. This compound, methyl tertiary-butyl ether (MTBE), reacted with the hydrocarbons in gasoline by donating an oxygen molecule that raised the fuel's octane level. This in turn made combustion more efficient and reduced the amount of greenhouse gases that vehicles emitted. As MTBE became more and more prevalent in gasoline, air quality began improving, smog began to subside, and cars produced exhaust containing less carbon monoxide. In California alone, smog-forming emissions from cars decreased by 15 percent within a year of introducing MTBE. Few people realized, however, that while air improved, MTBE had been leaking from gas station fuel tanks and trickling into groundwaters, many of which served as municipal drinking water sources. By 1998 aquatic ecologist John Reuter of the University of California–Davis stated in the *San Francisco Chronicle*, "When you find MTBE in so many drinking water sources throughout the state, it clearly indicates this is an issue of statewide and nationwide concern." A compound added to gasoline to improve environmental health showed the potential of causing an altogether new health threat.

Water utilities in other states soon detected the chemical in their drinking water and traced its source to gas storage tanks. The hazards versus the benefits of MTBE became a point of contention for environmentalists, who accused the oil industry of hiding information about MTBE's effects on environmental health. A 2005 press release from the Natural Resources Defense Council (NRDC) stated, "While the oil industry may not have known the extent of the MTBE health hazard, thousands of pages of internal industry documents and sworn depositions from recent litigation show that—without a doubt—the oil industry was well aware that MTBE would pollute drinking water." The oil industry grappled with the decision to reverse itself and begin formulating gas blends without MTBE.

MTBE has become a troublesome contaminant in water, but despite the NRDC's accusations, scientists have yet to determine the full effects of MTBE on human and ecosystem health. MTBE resists biodegradation, so it persists in the environment for several years, yet many states have resisted MTBE-gasoline bans because the compound helps state environmental agencies meet federal air quality requirements. Based on the premise that MTBE may be a health hazard, the EPA in 2000 banned its further use in gasoline. The EPA administrator at the time, Carol M. Browner, stated that, "Threats posed by MTBE to water supplies in many areas of the country are a growing concern. Action by Congress is the fastest and best way to address this problem."

She added, "Americans deserve both clean air and clean water—and never one at the expense of the other." Communities and the water treatment industry would be forced to address the MTBE problem.

Many states have had trouble complying with the need to convert all gasoline to MTBE-free formulas. The Energy Policy Act of 2005 has helped deal with this challenge somewhat by giving the EPA added authority to work with individual states in removing all gas station fuel tanks suspected of leaking MTBE into soils and groundwaters. As of 2008, however, the EPA lists on its Web site only 25 states that have partially or completely banned MTBE in gasoline.

Meanwhile, water quality experts continue to study the effects of MTBE on health and the environment. The American Water Works Association, the professional society for municipal water utilities, describes the MTBE dilemma on its Web site (URL: www.awwa.org): "MTBE is especially problematic because it has a low taste and odor threshold, tends to migrate in subsurface systems much faster than other constituents of gasoline, is difficult to remove from water at low concentrations via conventional treatment processes, and because of concerns regarding potential health impacts." Drinking water utilities have no legal requirements for eliminating MTBE from the water they treat, but the EPA asks that utilities keep MTBE levels under 40 parts per billion to avoid its bad taste and odor.

Even small changes in environmental laws have large effects on business. States such as California delayed complying with the EPA's ban for several years while fuel manufacturers devised new formulas using ethanol in place of MTBE. Businesses worried about higher costs of ethanol fuels and economists predicted a rise in fuel prices would affect large industries, thereby influencing world financial markets. MTBE remained in many gas blends for several years and continued leaking into the earth. Meanwhile, ethanol producers and farms producing corn, which is a major source of ethanol, looked forward to the ethanol-for-MTBE switch. There seemed to be no simple solution to fixing the MTBE problem.

In states that require MTBE-free fuel, change has come at a cost. Refineries spent large sums of money to convert their production to ethanol blends. In places such as California where emission standards are strict, per gallon gas prices are consistently the highest in the country.

Out-of-business gas station property contaminated with MTBE makes up a large portion of this nation's brownfields. Of the 425,000 to 600,000 brownfields in the United States, about half are underground storage tanks and petroleum products and a large portion of these sites contain MTBE. Fortunately, oxidation technology holds promise for removing this persistent chemical from the environment.

(continued from page 105)
to agrarian lifestyles in which industry and oil-burning machines are absent.

Real estate is an important facet of global economies. The potential value of land is determined by two economic factors: the risks involved in developing the land and the potential income from its development. Because redeveloping a brownfield site carries greater financial and environmental risks than redeveloping uncontaminated property, developers could lose a fortune if the project fails. This is why some brownfields sit untouched after purchase and why others become developed. The brownfield restoration projects under way today represent those with potential rewards that far outweigh potential risks.

A factory closes its doors and the building is demolished. A metal fence now encloses the empty grounds. All that remains are a few weeds here and there, a pile of cinder blocks, some half-buried storage drums. On hot, dry days, swirls of dust appear with each gust of wind. In rainy weather, puddles form, each pool covered with browns, greens, and purples. Most people see a blighted scene. Developers see an area of limitless possibility. People may not notice that the property is a short walk from a park that runs along a river. The ocean is within sight from a hillock. Though the property is ugly, the weather is almost always pleasant and downtown is five minutes away, lively with restaurants, a museum, and theater. From a developer's point of view, the property is far from being a wasteland.

Real estate development changes communities. Some people feel it helps contribute new homes and office space and new jobs. Others regret the influx of workers and shoppers, added traffic, and higher housing costs. Local government therefore weighs several issues before starting a brownfield project. First, a redeveloped brownfield site improves a community's aesthetic appeal. Few antidevelopment people would argue for keeping the contaminated site as is over some sort of cleanup and development, even if it is merely turned into an open field. Second, redevelopment attracts new businesses, which enrich tax coffers and perhaps the identity of the community. Emeryville, California, for example, lost much of its industry in the 1970s, resulting in hundreds of vacant acres containing contaminated soil and water. The city received a grant through the EPA's brownfield program and the land was purchased and redeveloped into one of the country's largest biotechnology centers. Third, redevelopment creates jobs, further boosting the local economy. Fourth, redevelopment creates new living space and workspace and therefore shorter drives for commuters. Finally, land that once threatened the health of the populace and food webs is decontaminated. From a purely ecological stance, cleaning up a brownfield is an undeniable plus.

The thousands of today's brownfields represent an array of land uses. Examples of the new uses for redeveloped brownfields are waterfront shops and restaurants, new parks and recreation areas, open space, urban community gardens, downtown office and parking space, downtown shopping districts, affordable housing in low-income areas, new sites for light industry, land for schools and universities, and restoration of wildlife habitat. Every state possesses brownfields; some states contain hundreds of known brownfields and an unknown number of privately managed brownfields. Some brownfields are large pieces of property and others are a fraction of an acre. In fact the number of the nation's brownfields exceeds the number of Superfund sites, and brownfields are far more diverse.

The EPA's Brownfields Action Agenda was created in 1995 to provide further help to communities planning a brownfield restoration. The program's four objectives are: (1) to provide seed money for getting projects under way; (2) to remove liability barriers that could stall redevelopment; (3) to open communications among local government, developers, and community residents during cleanup and redevelopment; and (4) to provide jobs and training for the local workforce during the project. In 1998 the EPA partnered with Argonne National Laboratory and the U.S. Army Corps of Engineers to create an information resource called the Brownfields and Land Revitalization Technology Support Center (or BTSC, for Brownfields Technology Support Center). The BTSC (URL: www.brownfieldstsc.org) offers a starting point for any group or individual planning a brownfield project.

REDEVELOPING BROWNFIELD SITES

The most rewarding part of brownfield cleanup is the day when the cleaned and restored property is again available for a community's use. The cleanup phase consists of all means of removing contamination from the brownfield, and this phase ends when the EPA issues a No Further Action letter. The restoration phase follows, and it is complete when the land has been returned to a safe condition suitable for use. In some instances, restoration consists of landscaping for a city park or the construction of new buildings. In other cases the community returns the restored land to natural conditions so that ecosystems can return. In 2008 the president of the U.S. Conference of Mayors, Mayor Douglas Palmer of Trenton, New Jersey, said in the group's annual report, "Brownfields are too costly to ignore, not only from the environmental standpoint of contamination, but also the social aspect of decayed properties and the potential they hold."

If a restored site is developed with buildings, professional property managers supervise the site's new structures. These managers oversee the property's finances and physical maintenance and they ensure that the land is not recontaminated. To do this, the manager may request periodic soil and water testing. In 1995 the EPA formed the Brownfields Economic Redevelopment Initiative, which guides landowners and managers on methods to monitor the site for any signs of new contamination. The Initiative also offers guidance on how communities can reuse the land in a sustainable manner.

The EPA further collaborates with brownfield projects outside the United States through its Office of International Affairs. Russia, Poland, the Czech Republic, Germany, the Netherlands, Canada, and England are among the countries that use U.S. brownfield restoration as a blueprint for their own environmentally damaged land. Germany in particular has led the way in developing a ranking system to be used in evaluating the economic benefits of a brownfield to be cleaned.

The cost of cleaning up contamination is often the greatest obstacle in the United States and abroad. In the United States alone, hundreds of brownfields sit idle because of a lack of funds even though the EPA supplies grants for much of the planning phase. Brownfields usually remain abandoned when the cleanup costs become greater than the money that can be earned from the restored site. This happens when the value of the land decreases during the time it takes to clean it. Environmental organizations have recently begun to arrange financing for brownfield cleanups and help make progress on the country's backlog of brownfields in order to avoid the wasting of land due solely to lack of money. Their goal is to return the land to natural conditions, such as wetlands, riparian areas, grasslands, or woodlands, rather than urban development, but any renewal of brownfields ultimately benefits the environment. Mayor Patrick McCrory of Charlotte, North Carolina, explained in 2008, "Many cities are experiencing a renaissance and residential booms in their downtowns and center city where there is a renewed interest and demand by residents to live closer to the hub of the city. Brownfield redevelopment can fuel this boom and is an essential component to the economic and environmental growth of a modern city. This is the only way to truly grow in a sustainable manner—save green space and redevelop land that was previously used and put it back into positive use."

CONCLUSION

A brownfield is an industrial or military site that has been abandoned or left idle and unused and is not being developed to its fullest potential because of contamination or perceived contamination. Brownfields can be cleaned up and redeveloped for use. This cleanup-to-reuse process requires careful planning with guidance from the EPA. It is accomplished by developers or new owners of the land on a voluntary basis. Each brownfield project consists of financial and environmental risk assessment, testing, and a written restoration plan. It ends with an EPA-issued No Further Action letter and the return of the land for use. Restored brownfields are used for urban redevelopment, parks, green space, or other sustainable activities. The EPA's Brownfield Program is unique because it includes economic issues along with environmental needs.

Brownfield cleanup uses all the available cleanup technologies currently available, including bioremediation and phytoremediation. Some communities prefer these natural systems for restoring the land, but other communities must work faster, either for financial reasons or health reasons, and they prefer excavation and other fast cleanup methods. Small brownfields may be completed in a few to several months, while large brownfield projects can take several years to complete.

Estimates on the number of brownfields currently in the United States reach as high as 500,000 or more. Many of them have been or will be left unfinished because of the expense of the cleanup or the time required to complete the cleanup. There are success stories, however, such as Love Canal in New York. Love Canal was the first incident of severe pollution that had been recognized by the government as an environmental disaster. It also became one of the first qualified successes in the U.S. brownfield program. Despite many challenges, this program offers the best prospects for returning environmentally damaged land to a safe and reusable condition.

Remediating the Water Supply

The evolution of life on Earth and the birth of civilization are intertwined with the purity of the world's water. Societies could never have formed communities, developed commerce, or built stable governments were it not for a safe and clean supply of drinking water and access to the sea for transporting goods. All of Earth's ecosystems depend on water to sustain life and drive the planet's *biogeochemical cycles.* The decrease in water resources due to pollution is therefore a critical environmental concern. For these reasons civilization's relationship with water is reviewed here.

The Earth's biota cannot exist without the oceans and freshwaters. Nearly three-quarters of Earth are covered by oceans, seas, and bays, contributing to about 97 percent of its surface water. Oceans dictate climate and weather patterns, control most of the planet's respiratory system and temperature, and store energy in the movement of tides. The oceans also hold more aquatic ecosystems than anywhere else and probably contain many more living things and ecosystems that science has not yet discovered. Lakes and streams have provided the earliest settlements with water for drinking and cooking, hygiene, and ceremonial events. Rivers helped establish the first agricultural communities. Primitive commerce routes began on rivers flowing to neighboring communities, and with travel, humans' knowledge about the world grew.

The oceans and the atmosphere are fluid media, meaning their conditions are ever-shifting. The ocean interacts with the atmosphere in two ways, one physical and one chemical. First, oceans physically exchange

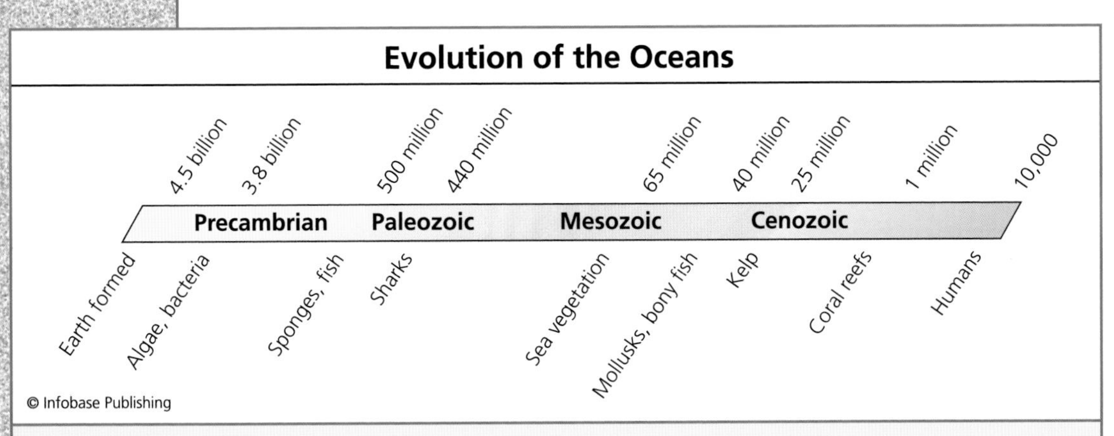

Life on Earth began in the ocean, and the oceans have always contributed to the planet's natural rhythms. Only since humans emerged on Earth have marine ecosystems become threatened. Some of the damage may be irreversible.

heat, water, and tidal momentum with the atmosphere. As Earth matured, currents carried the Sun's heat energy, absorbed mainly at the equator, toward the poles. Cold air then escaped into space and the planet cooled sufficiently to support life. Currents made nutrients available for the evolution of ocean biota. Ecosystems on land developed and adapted to air temperature gradients and winds powered by the currents. Second, much of the heat escaping from oceans does so by evaporation. Water vapor rises into the atmosphere and chemically reacts with compounds there to form the greenhouse effect, which results in a warm protective blanket for life on the Earth's surface.

> Once, long ago, Sun was the ruler of all the Earth. Next to him, the other spirits were as the sparrow beside the grizzly bear. So the spirits had a secret meeting and elected the water-spirit to approach the Sun to give up some of his power.
>
> Water went to Sun, and formed a clear, deep pool at his feet. When Sun saw his own face reflected in the pool, he was so delighted that he promised Water anything she wanted. When she demanded some of his power, he realized that he had been tricked, but according to his word, he gave power to all of the other spirits. Water, for her part, got more than anyone, and became, next to Sun, the most powerful force on Earth.
>
> —*Cree Indian Legend*

This chapter explores the critical role aqueous environments (those containing mainly water) have on Earth and the methods used for cleaning up contaminated waters. The Earth contains about 330 million cubic miles (1.385 billion km^3) of water of which almost 97 percent consists of oceans, seas, and bays. These marine waters are 3.5 percent salt (3.5 grams of salt dissolved in 100 grams of water) plus traces of all the elements found on Earth. Saline groundwaters and saline lakes also contribute a small amount to the total salt waters. Nonsaline, or freshwater, makes up only 3 percent of the planet's water. Most of the world's freshwater is stored in the polar ice caps and glaciers (about 70 percent) and in groundwater (about 30 percent). Much smaller amounts are located, in decreasing order, in permafrost, lakes, soil, the atmosphere, swamps, rivers, and biota.

Surface waters and groundwaters both participate in the Earth's *water cycle,* also called the hydrologic cycle. The world's surface waters, especially oceans, and to a smaller extent groundwaters also play a part in biogeochemical cycles. Environmental scientists study biogeochemical cycles as a way to keep track of the supply and depletion of many of Earth's elements. Nitrogen, sulfur, phosphorus, potassium, calcium, sodium, hydrogen, and carbon would not be available in the forms needed by living systems without chemical conversions within their cycles. At least part of every biogeochemical cycle takes place in an aqueous environment.

In the carbon cycle, marine algae and plant life capture a portion of atmospheric carbon dioxide during photosynthesis, and marine animal life exhale carbon dioxide. Meanwhile, carbon is stored in seawater and sediments as biota die and decompose. In nitrogen cycling, leachates containing nitrates (NO_3^{-1}) and nitrites (NO_2^{-1}) enter water where bacteria such as *Pseudomonas* denitrify them to nitrogen gas (N_2). Water bacteria such as *Acetobacter* and blue-green algae convert the gas into nitrate and nitrite compounds (nitrogen fixation), which are in turn taken in by nitrifying bacteria living on the roots of plants and trees. *Nitrosomonas* and *Nitrobacter* bacteria supply nitrogen to plants in this process called *nitrification.* Nitrogen eventually returns to waters after dead plant matter decomposes. Phosphorus, sodium, sulfur, and other elements cycle into and out of the oceans in similar ways.

Since life depends on these physical actions and chemical reactions, maintaining clean and healthy aquatic environments is the world's priority. The World Health Organization (WHO) and UNICEF estimated in their joint 2005 report titled "Water for Life" that a person uses five gal-

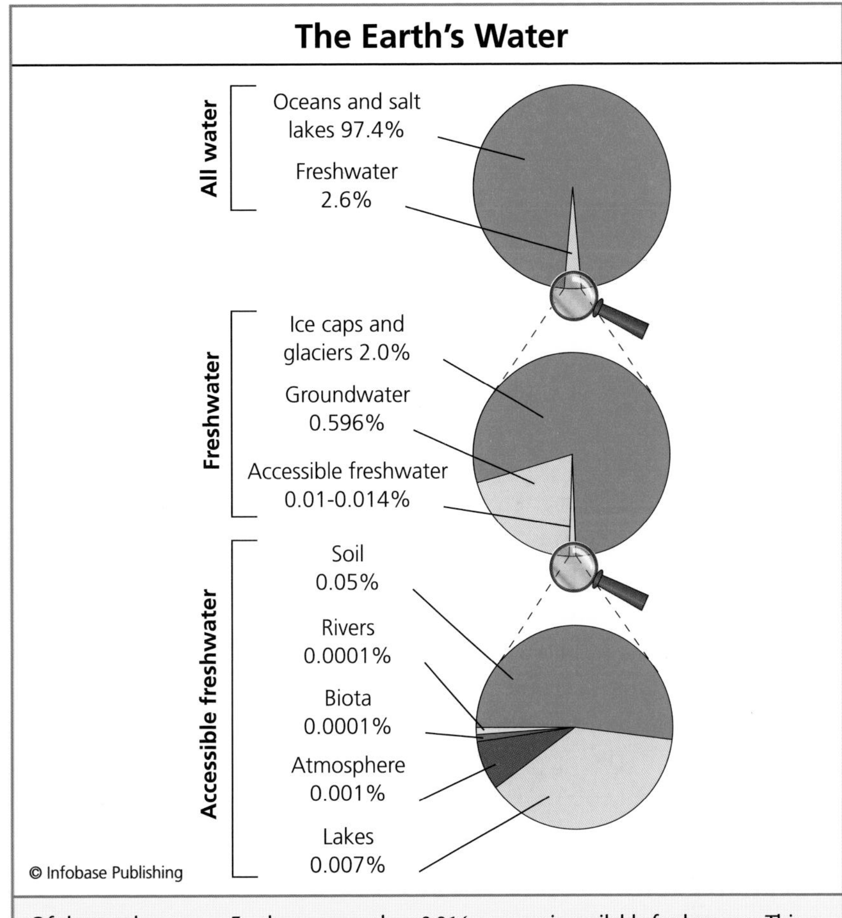

Of the total water on Earth, no more than 0.014 percent is available for humans. This accessible water is used for drinking, household use, some manufacturing processes, and recreational uses.

lons (20 l) of water per day. Despite what seems to be a massive amount of global water, a very small portion of it is accessible for daily use. As populations increase, they use an increasing percentage of the planet's surface runoff. Today humans draw about 34 percent of runoff for their immediate needs and use another 20 percent for fishing, boating, and diluting wastes. In some parts of the world the withdrawal from surface water is much higher.

In 1854, a swift-moving cholera outbreak in London's Soho district killed 600 people. Physician John Snow seemed more determined than many of his colleagues to find the root of the epidemic as he struggled to treat an influx of patients. He questioned patrons of local pubs and hotels and the residents of the surrounding community, focusing on their daily water supply among other habits. Most people used community pumps fed by London's water distribution pipes. Following his instincts, Snow identified one pump on Broad Street serving as a common water source for a large portion of the community. Those who used their own wells avoided getting sick, but those using the Broad Street pump often fell ill. Other doctors were hesitant to accept Snow's findings, especially since he had never identified a chemical or biological cause of the disease. Yet his meticulous collection of evidence and his logical process of elimination won over doubters. An investigation showed that the suspect pump drew from a well that had been drilled to within a few feet of a corroded sewer line. London's cholera epidemic was halted by the simple step of removing the pump's handle. John Snow's investigation made him the first person to connect an illness with water contamination.

From the 19th century onward in the United States, water pollution caused troubles in a variety of places and circumstances. California's ore mining industry provides one example. Waters flowing from mining sites were so tainted with metals and acids that crops irrigated with them soon died. The state in 1884 passed a law prohibiting the use of water flowing from the mines, but this of course this did not stop the cause of the pollution. Throughout the United States and Europe, waters would become more and more contaminated with pesticides and herbicides, animal waste, and industrial discharge.

Industries and municipalities had long used rivers as the best mode of carrying their wastes from sight. People assumed salt waters and various marine creatures would digest all matter once the wastes flowed into the ocean. Before the earliest laws against water pollution, pollution was viewed merely as an inconvenience. Bad odors emanating from rivers and lakes and the occasional piece of refuse bobbing on the waves were unsightly but not thought to be dangerous. Rachel Carson, Jacques-Yves Cousteau, and a new generation of environmentalists soon began alerting the public to a growing danger in the water, and by 1972 scientists estimated that only a third of U.S. rivers were safe for swimming or fishing.

The 1972 Federal Water Pollution Control Act was America's first attempt to control how open waters were used. Many leaders at the time held skeptical views on the need for strong environmental laws. In October 17, 1972, President Richard Nixon cited his opposition to the costs of the proposed law and vetoed it. In his message to Congress he rationalized, "The pollution of our rivers, lakes and streams degrades the quality of American life. Cleaning up the nation's waterways is a matter of urgent concern to me, as evidenced by the nearly tenfold increase in my budget for this purpose during the past four years. I am also concerned, however, that we attack pollution in a way that does not ignore other very real threats to the quality of life, such as spiraling prices and increasingly onerous taxes. Legislation which would continue our efforts to raise water quality, but which would do so through extreme and needless overspending, does not serve the public interest. There is a much better way to get this job done." Despite Nixon's misgivings, U.S. Congress overrode the veto the very next day. The new law dealt with the known threats that had been coming from industrial discharges, but as populations reached outward from urban areas, the new law had merely a modest effect on total pollution. Municipal wastes replaced industrial discharge in many areas as the main source of water pollution. Amendments to the act began to address these forms of pollution too, and in 1977, the U.S. Congress passed the Clean Water Act.

The health of the world's waters outside the United States and Canada is often far more endangered. The WHO estimates that 1.8 million deaths each year are due to gastrointestinal illnesses from pathogens in water, especially in young children. Many developing nations possess poor water-supply infrastructure and their industries and cities continue to pour wastes into local waters. War and threat of genocide contributes by driving millions of people to areas where water and food are in short supply. Countries in the Middle East that draw water from the Nile River, such as Sudan, Ethiopia, and Egypt, are thought by many political experts to be under the greatest threat of a war over water. These regions are already experiencing conflicts and genocide. Countries along the Jordan River—Jordan, Israel, Syria, and Lebanon—also suffer from cultural strife made worse by water shortages. To make matters worse, many of these countries have the world's highest population growth rates in the world. The World Bank and the WHO estimate that between 1.1 and 1.4 billion people in developing countries have inadequate access to water. Further pollution of all global water supplies should not continue even another day, and yet it does.

HOW OUR WATER BECAME POLLUTED

Water pollution is any chemical, physical, or biological change in water quality that makes it unusable or causes harm to ecosystems. Pollution can occur naturally due to seismic shifts or violent storms, but most water pollution today is because of human activities. Surface water pollution comes from the air, land, and from tributaries, and may take the form of chemicals, biological matter, or physical changes in the water itself. Chemical pollution includes inorganic and organic substances. Pesticides, industrial chemicals, household cleaning products, metals, paints, and radioactive compounds are some of the many chemicals that can dissolve or suspend in water. Biological pollutants consist of microscopic bacteria and viruses from raw sewage, protozoa, and algae or macroscopic parasites, worms, and shellfish. Lakes overrun by invasive plant or fish species may also be thought of as biologically polluted. Physical pollution of water is due to soil and silt carried in runoff. These particles block sunlight from reaching photosynthetic plant life. Another physical force is evaporation. Prolonged drought prevents replenishment of the water lost through evaporation, and salt and metal concentrations in the remainder rise to toxic levels. Toxic waters stunt plant growth and cause deformities in plants and animals. Finally, thermal pollution represents a type of physical pollution caused by the heating of water, explained in the sidebar "Thermal Pollution of Water" on page 126. Equipment-cooling water used in industrial plants and water from electric power plants are warmer than aquatic life can withstand, so they die and ecosystems suffer.

Water pollution may begin at a single source or a variety of sources at the same time. *Point sources* are those that disgorge pollutants at specific locations. Drainpipes from factories, broken sewer lines and oil pipelines, wastewater treatment plants, ports, leaky hazardous-waste burial sites (deep landfills, pits, trenches, or abandoned mines), and dumped wastes (batteries, scuttled boats, fishing lines) are point sources. Oil tankers that leak their cargo or run aground are also point sources even though they are mobile.

Pollutants from *nonpoint sources* are carried in rainfall, erosion, and runoff. Agricultural lands, manufacturing plants, forests, lawns and golf courses, parking lots, and service stations are some of the origins of this type of pollution. Rainfall may carry acid (acid rain) and snow carries particles to surface waters. Hazardous substances also percolate through soils until they reach aquifers. Some aquifers are unconfined, which means they are not protected by surrounding bedrock and therefore connect to an underground water table. Polluted unconfined aquifers can feed lakes or rivers below their surface, and the contamination then spreads to much larger areas. Diverse human activities and natural occurrences have a profound effect on water quality. The categories of water pollutants described in the table on page 120 affect not only human health but all biota and biogeochemical cycles.

Water presents a problem compared with pollution on land: It moves. Pollutants in soil also migrate but the movement takes place slowly. Fast-flowing rivers differ because they disperse over distances quickly. Pollutants may also leach from wastes that have been discarded into bodies of water and fall to the bottom. For example, lead leaches out of a number of items when they are tossed into water, including corroded car batteries, fishing weights, diving weights, ammunition, paints, piping and fixtures, and even some ceramic dishes.

Water pollution did not arise with the birth of industry. People heaved garbage into the nearest stream centuries ago. One hallmark of Europe's Middle Ages was a deplorable lack in hygiene practices. People dumped household garbage, sewage, and dead animals into streets to be washed away with the rains. Until the rain came, the trash provided food for vermin. As a consequence, bubonic plague epidemics throughout the Middle Ages decimated the population. In 1388 the British Parliament banned the use of waterways for the wastes filling London's Thames River. New Amsterdam (now New York City) enacted a similar ban in 1657 and again in 1866 with little success. Rivers and the sea remained society's disposal site despite such bans. In the late 1800s New York City resorted to constructing a platform to jut over the East River to give residents an authorized place to hurl their garbage into the flow!

So much refuse clogged waterways in European and American cities in the 1800s that tradesman could not get their boats up and down river. Congress passed the Rivers and Harbors Act in 1899 to help alleviate the problem but not until 1948 did the United States address the actual cleanliness of its water. Almost a century after John Snow's efforts, the U.S. Congress passed the Water Pollution Control Act to reduce pollution in interstate waterways. Congress renamed it the Clean Water Act in 1972 and placed the Environmental Protection Agency (EPA) in charge of enforcing it. The EPA now monitors the pollutants discharged into water

(continues on page 122)

THE MAIN CATEGORIES OF WATER POLLUTANTS

CATEGORY	EXAMPLES	SOURCES	EFFECTS	PRINCIPAL WATER AFFECTED
organic chemicals	oil, fuels, pesticides, solvents, detergents, plastics	industrial discharge, households, surface runoff	nerve damage, cancer, reproductive disorders	fresh surface, groundwaters, coastal marine, oceans
inorganic chemicals	toxic metals, salts, acids	surface runoff, industrial discharge, households, acid rain, salt water infusion	diseases (cancer, liver disease), nerve damage, contaminated drinking water	fresh surface, groundwaters, coastal marine
sediments	soil, silt	erosion	clouds water and reduces photosynthesis, carries pesticides, disrupts food webs	fresh surface, coastal marine
metabolism disruptors	organic wastes high in nitrogen or phosphorus			
	manure, plant debris	sewage, feedlots, food manufacturers, pulp mills	blooms, eutrophication, fish kills	fresh surface, coastal marine

The Main Categories of Water Pollutants				
Category	**Examples**	**Sources**	**Effects**	**Principal Water Affected**
	fertilizers	agriculture, residences	blooms, eutrophication, fish kills	fresh surface, coastal marine
	endocrine disruptors	hormones, organic compounds, PCBs	reproductive disorders, birth defects, damaged immune systems	fresh surface, groundwaters, coastal marine, oceans
infectious agents	bacteria, viruses, protozoa, worms	sewage and animal wastes	disease	fresh surface, groundwaters, coastal marine
radioactive materials	isotopes of uranium, cesium, iodine, and radon	nuclear and coal-burning power plants, nuclear weapons production, mining	mutations, cancers, reproductive disorders, birth defects	fresh surface, groundwaters
heat	thermal pollution	cooling waters from electric power and manufacturing plants	thermal shock in fish leading to death or increased likelihood of disease	fresh surface, coastal marine

(continued from page 119)
and sets water quality standards. Thanks to advances in sensitive analytical techniques, scientists can detect minute amounts of chemicals and run hundreds of samples in just a few hours. In 1972 less than 40 percent of U.S. waters met the Clean Water Act's requirements. In 20 years the figure had improved to 60 percent and it has remained near this level ever since.

Other parts of the world do not have the advantages of western societies, and their waters have suffered because of it. Municipal wastes plus decayed sewer pipes and water distribution lines threaten health in many countries. India, for example, has a strong economy yet a weak infrastructure for utilities. Over one-twentieth of the world's population lives in the watersheds of India's Yamuna and Ganges rivers, which daily receive municipal and industrial effluent—the Ganges may receive up to a quarter of a billion gallons (946 million l) of sewage every day. Yet millions of people hold these waters sacred and depend on them for good fortune and immortality. Each year worshippers walk into the Yamuna and the Ganges during the Hindu Ardh Kumbh Mela festival to honor the river waters. Unfortunately, these waterways carry cholera, typhoid, and dysentery. New Delhi's Centre for Science and the Environment monitors India's water pollution and tries to educate communities and industries on the harm wastes cause in the country's lifeblood. But advanced treatment technologies are too expensive for most communities that rely on rudimentary treatment systems or have none at all. Still, India's residents have demanded their government do something about the country's water pollution. According to a March 18, 2007, *Los Angeles Times* article on India's pollution, monastery leader Narendranand Saraswati said, "The government has promised us they would stop dirty water flowing into Mother Ganges, but it's still being done. We want the entire country to know we will not stop until the river is clean!"

In North America, agriculture, industry, and surface mining represent the main sources of pollution. The EPA identifies agricultural nonpoint sources as the major polluter of inland lakes and rivers. In the agency's most recent (2002) assessment, "National Water Quality Inventory: Report to Congress," agricultural pollution affected almost 40 percent of the 113,663 miles (182,923 km) of rivers and streams EPA scientists sampled. Agriculture contributes the following materials to surface waters: sediments from overgrazed rangeland, fallow (uncultivated) cropland, and erosion carrying fertilizers, pesticides, herbicides, manure, and feedlot and slaughterhouse wastes.

Organic wastes such as yard trimmings and grass clippings are often included as part of agricultural waste. High levels of organic wastes lead to hazardous conditions best illustrated by the dead sea that forms each year in the Gulf of Mexico. Farm runoff in the Mississippi River enters the gulf, bringing enormous amounts of nitrogen and phosphorus compounds. These elements are normally in limited supply in natural ecosystems. When large amounts wash into waters, algae in the water respond by undergoing rapid growth, called an algal *bloom*, due to the new and available nutrient supply. As the algal cells die, bacteria devour them and a bacterial bloom ensues. The bacterial bloom consumes all of the dissolved oxygen in the surrounding water and the region soon becomes uninhabitable for fish and other marine life. The entire process is called *eutrophication*. Many such dead spots plague marine ecosystems around the world each year. In the United States, the EPA and the Department of Agriculture are working to solve water pollution from agriculture.

Household chemicals contribute to pollution in rivers and bays in developed countries. An expanding list of chemicals possibly toxic to

Household Products Suspected of Polluting Water

Ingredient	Products
antibacterial triclosan	deodorants, toothpastes, hand soaps
perfume fragrances	perfumes, cosmetics, shampoos
dyes	food and nonfood product packaging
dibutyl phthalate	nail polish
polyvinyl chloride	food wrapping, shower curtains, garden hoses
bisphenol A	buckets, pails, pet dishes, bottles, other hard plastics

Note: Information according to the 2007 report "Down the Drain," published by the Environmental Working Group. Available online. URL: www.ewg.org/reports/downthedrain. Accessed September 19, 2008.

aquatic life is now found in fresh and salt waters, as shown in the table on page 123. Equally disturbing, water treatments may not completely remove these compounds.

Endocrine disruptors (EDs) are a group of compounds hazardous to ecosystems and released from various household and industrial substances. Some known EDs are the drug diethylstilbestrol, dioxins, PCBs, some pesticides, and certain compounds found in detergents and plastics. EDs dissolve in water and are ingested or absorbed through animal skin. The absorbed compounds then interfere with the endocrine system, which makes and regulates reproductive hormones. EDs affect humans as well as aquatic species and environmental scientists continue to learn more on the effects of EDs on ecosystems. To date, some of their findings include the following:

- marine snails, turtles, and fish nearing extinction due to disruption of female reproductive systems
- distorted sex organ development in American alligators
- reproductive and immune systems damaged in polar bears, mink, rabbits, Baltic grey seals, ringed seals, and harbor seals

Biologist Tim Verslycke of the Woods Hole Oceanographic Institution in Massachusetts proposed in 2007 in the institution's *Oceanus* magazine that ED research begin with aquatic invertebrates. These organisms form the foundation of aquatic food chains so the effect of EDs on their metabolism relates to larger food webs. But there is a daunting variety of invertebrates from which to select for study. Verslycke and others are now setting up study models using different invertebrates to examine the effects of EDs on metamorphosis, reproductive cycles, molting, and hormones. He explained, "Researchers are now finding that some chemicals, especially pesticides, cause unexpected and unintentional harm to many invertebrate species that play essential roles in marine ecosystems and food webs."

Biological pollution ranges from single molecules to large organisms. For instance, medically prescribed antibiotics flow from homes through sewers and on to wastewater treatment plants. Refined equipment now detects very low concentrations of antibiotics and other drugs in municipal water, suggesting that treatment plants do not remove all of these compounds. In 2008 the *New York Times* interviewed Paul Rush, deputy commissioner of New York City's Environmental protection. "A person would have to drink 1 million glasses of water to get the dose of even one over-the-counter ibuprofen tablet or the caffeine in one cup of coffee," Rush assured. "Even at eight glasses of water per day, this would take the average person over 300 years to consume." The ability of small amounts of drugs to endanger other organisms may be another matter. Antibiotics interfere with the good bacteria needed for decomposing sewage organic matter and probably cause further trouble when discharged into the environment. On a bigger scale, oceangoing ships have been blamed for bringing alien aquatic animals and plants into new ecosystems. The zebra mussel, *Dreissena polymorpha,* is an infamous example of such an aquatic invader. This bivalve mollusk—more commonly known as a clam—is thought to have originated

Zebra mussels, about one half-inch (1.3 cm) across, are invasive species in the Great Lakes and represent a form of biological water pollution. The mussels form dense populations that clog water utilities' intake screens, foul the bottom of boats, and displace native species in lake ecosystems. *(ECHO Lake Aquarium and Science Center)*

THERMAL POLLUTION OF WATER

Industrial and electric power plants withdraw water from lakes and rivers and use it to cool equipment heated in normal operations. The plants then return the warmed water to their original sources. This common practice causes thermal pollution, which is a warming of naturally cool waters to higher temperatures. Aquatic ecosystems evolved to use certain temperature ranges and they carry out their metabolism within these ranges. Warm or hot water discharges into cool waters raise temperatures as much as 35°F (2°C), and though this change is temporary, aquatic life cannot adapt to it and suffers thermal shock. Organisms that do not die often experience other long-term damage, such as reduced resistance to disease, parasites, and toxic chemicals.

Environmentalists have long argued that using less energy ultimately leads to less thermal pollution of water. A 2002 press release from the Riverkeeper environmental organization that monitors New York's rivers stated, "The truth is this ecological damage to the Hudson River is unnecessary. The power plants use 'once-through' cooling, the most wasteful and destructive system. By contrast, common 'closed-cycle' technology used on the Hudson's new power plants . . . could reduce water withdrawals and fish mortality by up to 97 percent."

As Riverkeeper suggests, three alternatives exist for preventing thermal pollution. First, facilities can discharge heated water into artificial ponds and canals. The cooled water can be either reused or returned to the environment. Second, heat may be redirected into the atmosphere by cooling towers. Wet cooling towers allow hot water to cool within an upward air stream until the moisture exits into the air. Dry cooling towers, by contrast, confine the hot water in pipes and cool it by air circulation, after which it leaves as dry air. In the third option, heated water is turned to steam with a small energy input, and then enters a turbine. The turbine converts steam energy to motion, which runs an electric generator. Conversion of thermal energy to motion energy follows Newton's third law of physics, which states "to any action, there is an opposite and equal reaction." Thermal pollution represents a large loss of potentially usable energy. Environmental engineers will someday tap into this renewable energy source.

in the Caspian Sea and spread to European ports via inland waterways in the 18th century. Freighters discharging ballast water are believed to have introduced it to North America, probably in Lake St. Clair in the 1980s. By the end of the 1980s the mussels had spread to all of the Great Lakes and clogged intake and outflow pipes, interfered with pumping stations, and jammed boat motors. They have brought an ironic benefit, however, by serving as food for migrating bird species that have suffered declining numbers in the past decade. One thing is certain: Zebra mussels cannot be ignored because they produce 30,000 to 40,000 eggs per season.

COMMERCE VERSUS ECOLOGY

The time line titled "Development of Ocean Commerce and Exploration" on page 128 illustrates the use of water in history for commerce, food, and knowledge. Today global shipping of products and the fishing industry put the greatest demands on the world's oceans. Global cargo shipments rely on oceangoing ships to transport billions of dollars' worth of commodities each year, but at the same time these ships contribute to ocean pollution and the movement of invasive species. The world's fishing fleets, meanwhile, have hunted dwindling stocks to supply 6 percent of the population's protein needs. In Asia fish supplies 30 percent of dietary protein for nearly 1 billion people. (The Monterey Bay Aquarium in California estimates nearly half of the total catch now comes from fish farms.) Humans eat about 1,000 species of the 30,000 species of fish on Earth, but just a few species dominate within that thousand: Alaskan pollack, Peruvian anchovy, Atlantic bluefin tuna, and Chilean jack mackerel. In the United States and other wealthy countries, grocery stores have supplied an increasing variety of seafood and this may give the public a false impression that fishing is a prosperous industry. The variety may in fact be telling the world that many previously common catches have decreased to very low levels. The United States has leveled off in its yearly catch since 1985, but catches in other countries with economies tied to fishing have plummeted. The fishing industry particularly supports economies in developing countries where 90 percent of the world's fishing-related jobs occur. Overfishing, pollution, and warming ocean temperatures at their current rates may eliminate the world's ocean fisheries by 2050, an event that has the potential to push these countries toward disaster.

DEVELOPMENT OF OCEAN COMMERCE AND EXPLORATION	
YEAR	
25,000 B.C.E.	Humans migrate over waters by vessel
3500	Egyptians use boats for trade in Mediterranean
3000	Greeks sail outside the Strait of Gibraltar
1600	Phoenicians sail around southern tip of Africa
1500	Early navigation equipment
900	Vikings reach Iceland and Greenland
50 C.E.	Nero's Rome receives grain fleets from Alexandria
1400s	First nautical charts and mariner's compass
1492	Columbian Exchange trade route opens between North America and Europe
1769	Benjamin Franklin draws the first chart of the Gulf Stream
1842	Charles Darwin publishes his findings on coral reefs
1849	Discovery of continental shelf
1853	Life found at depths greater than 1,000 fathoms
1872	Challenger global expedition begins; runs through 1876
1882	USS *Albatross* becomes first vessel for marine research
1930	Woods Hole Oceanographic Institution begins ocean studies
1950	World fish harvest is 19.3 million tons (17.5 million metric tons)
1964	First deep-ocean submersible, *Alvin*, is launched
1977	Discovery of hydrothermal vents at sea bottom
1988	Japan's ocean, coastal, inland fishing totals 12.5 million tons (11.3 million metric tons)
2001	U.S. ship freight is 1.3 billion tons (1.2 billion metric tons), $718 billion
2003	World fish harvest is 132.5 million tons (120 million metric tons)
2005	Japan's ocean, coastal, and inland fishing totals 6 million tons (5.4 million metric tons)
2008	Ships supply 99 percent of U.S. imported raw materials
2010	Acoustic modeling of oceans begins counting and classifying organisms as small as plankton

Environmentalists have become alarmed over the state of the oceans' fish populations, especially species that do not directly support a country's economy. For example, 150 million sharks are killed each year for fins and other body parts that make their way into food, cosmetics, apparel, and medicines, but none of these products plays a vital role in industry. The international Shark Research Institute estimates that more than 100 million sharks are slaughtered annually; perhaps over 70 percent of the world's shark population has been lost. Shark specialists lack actual numbers on shark populations because these animals are difficult to study and few countries have methods to monitor shark numbers. Currently only Canada, Australia, New Zealand, and the United States have begun to monitor and manage shark populations.

International efforts for protecting economies based on fishing as well as aquatic life have focused on several priorities. These are as follows:

- maintaining profitable shipping industries, ports, and harbors
- supplying safe drinking water by maintaining or building infrastructure
- providing clean water for recreation
- managing watersheds and coastal zones for human and animal use
- encouraging national laws for protecting waters and aquatic life, and
- developing sustainable fishing and clean (nonpolluting) cargo ships

FRESHWATERS

Freshwater accessible for human use makes up only 0.014 percent of the total water on Earth. It is one of the world's most precious commodities, and any reduction in supplies can have devastating effects on community health. Agriculture and manufacturing use most of the freshwater. Large and growing populations also demand large amounts for drinking water and other uses. After humans take their share of water for these uses, wildlife and natural plant life must exist on the remainder.

Reliable runoff is surface water that supplies a stable source of clean water to meet human daily requirements. Because humans depend on reliable runoff above all other sources, it must be available and dependable from year to year. Freshwater supply is threatened by global warming, which leads to too little water (droughts), too much water (floods), and pollution that can arise from either condition. Put another way, dry regions will get drier and wet regions will get wetter if global warming continues. In a 2008 article in *National Geographic* magazine, climate modeling expert Isaac Held of Princeton University summed it up in blunt style. He said, "There's nothing subtle here. Why do we need climate models to tell us that? Well, we really don't."

Receding surface waters become vulnerable also to saltwater intrusion from marine waters. Saltwater intrusion is becoming a dangerous form of contamination for both drinking water sources and fragile ecosystems throughout the world. People currently use about 54 percent of the planet's reliable runoff and at the current population growth the number may reach 90 percent by the year 2025. A water crisis looms. Held spoke on the subject in 2007 in Paris at the Intergovernmental Panel on Climate Change: "I am most concerned about the changes in tropical rainfall patterns that will accompany global warming, especially increased frequency and severity of droughts in underdeveloped regions that have the fewest resources to adapt to changing conditions. This is a question of 'environmental justice' with many regions of the developing world likely to be affected adversely, not due to their own actions, but as a consequence of the emissions of greenhouse gases by the developed world."

The main sources of reliable runoff pollution differ between developing and developed countries. In developing countries, polluted streams are oftentimes caused by poor or absent sewer systems. Infectious disease is one of the major causes of mortality in developing countries and much of the transmission occurs via waterborne routes. In developed countries, urban and agricultural point and nonpoint sources contribute chemicals, oils, and untreated wastes. Poorly maintained septic tanks and campsites close to lakes and streams also add untreated fecal matter.

Fast-flowing waterways are agitated (mixed vertically) and therefore aerated. The oxygen helps bacteria degrade organic matter. Flowing water cannot handle all the pollution, however, and high levels of organic compounds, salts, and metals simply flow downstream. International environmental organizations have suggested that all waters on Earth now contain some form of pollution and at least 500 of the world's rivers are heavily polluted.

Lakes differ from waterways by their lack of agitation and they contain stratified layers with slow flow rates. With little mixing, less dissolved oxygen is available for aerobic bacteria and little breakdown of organic matter can occur. Anaerobic organisms help, but their metabolism runs slower than aerobic reactions so anaerobic species do not remove pollutants quickly from the environment. Many toxic substances that enter lakes simply move toward the lake bottom and stay there. Pesticides, heavy oils, and metals harm the bottom fish, aquatic plants, and diving ducks. Raptors and mammals that feed on these species are harmed next, and before long, an entire food web begins to die. Lakes can also undergo eutrophication after receiving runoff high in fertilizers or animal wastes. Surface runoff adds road deicing chemicals, mine tailings, pesticides, and herbicides. Natural degradation systems remove some of these pollutants,

Despite the U.S. Clean Water Act, more than half of major manufacturing facilities in the United States annually exceed the legal limit of materials discharged into water. Water pollution also consists of constituents that are never measured: dumping from private property; accidental or intentional discharges from small businesses; and surface runoff. *(Plateau Software)*

THE PRESENT WATER STATUS OF SOME OF THE WORLD'S COUNTRIES
SEVERE WATER STRESS—MORE THAN 90 PERCENT OF LAND IN WATER STRESS
Israel, Trinidad and Tobago, Syria, Nepal, Kuwait, Azerbaijan, Belgium, Tajikistan, Kyrgyzstan, Turkmenistan, Macedonia, Tunisia
HIGH WATER STRESS—MORE THAN 75 PERCENT OF LAND IN WATER STRESS
Saudi Arabia, Egypt, Iran, Uzbekistan, Iraq, Armenia, Libya, Jordan, Lebanon, Morocco, India, Pakistan
MODERATE-TO-HIGH WATER STRESS—35 TO 75 PERCENT OF LAND IN WATER STRESS
United Arab Emirates, Spain, Algeria, South Africa, Turkey, Kazakhstan, Greece, Portugal, South Korea, Oman, Bulgaria, China, Mexico, Chile, Niger, Sri Lanka, the Netherlands
MODERATE WATER STRESS—15 TO 34 PERCENT OF LAND IN WATER STRESS
United States, Sudan, Somalia, Italy, Ethiopia, Cuba, Peru, Argentina, Bangladesh, United Kingdom, Albania, France, Namibia, Nigeria, Ukraine, Zimbabwe

but without cleanup, persistent compounds remain in lake sediments and riverbeds from years to decades.

The United States has adequate overall water supplies distributed across water-rich and water-poor areas. The eastern half of the United States almost always receives enough annual rainfall to replenish reliable runoff sources: 32–48 inches (81–122 cm) per year. The states west of the central plain region, however, receive half of that or less. Areas in Arizona, New Mexico, California, and along the Texas coast are experiencing conflicts between farmers, ranchers, and the fishing industry pitted against each other and against urban users, recreation sites, and the needs of wildlife.

THE PRESENT WATER STATUS OF SOME OF THE WORLD'S COUNTRIES *(continued)*
LOW-TO-MODERATE WATER STRESS—5 TO 14 PERCENT OF LAND IN WATER STRESS
Botswana, Bolivia, Mozambique, Philippines, Japan, Mongolia, Australia, Denmark, Mauritania, Moldova, Senegal
LOW WATER STRESS—LESS THAN 5 PERCENT OF LAND IN SOME WATER STRESS
Dominica, Russia, Vietnam, North Korea, Mali, Venezuela, Chad, Finland, Romania, Madagascar, Malaysia, Indonesia, Ecuador, Kenya, Germany, Colombia, Canada, Thailand, Sweden, Lithuania, Norway, Nicaragua, Brazil, Estonia
NO WATER STRESS AT PRESENT
Angola, Austria, Belarus, Benin, Bosnia and Herzegovina, Burkina Faso, Burma, Burundi, Cambodia, Cameroon, Costa Rica, Croatia, Czech Republic, Democratic Republic of Congo, El Salvador, Ghana, Guatemala, Guinea, Haiti, Honduras, Hungary, Iceland, Ireland, Laos, Latvia, Liberia, New Zealand, Panama, Papua New Guinea, Paraguay, Poland, Rwanda, Sierra Leone, Slovakia, Slovenia, Switzerland, Tanzania, Togo, Uganda, Uruguay, Zambia
Note: countries are listed within each category from most stressed to least stressed except for the "No Water Stress at Present" category in which countries are in alphabetical order.

In other parts of the world, water depletion is an even greater crisis. A belt can be drawn around the globe to include countries in North America, northern Africa, the Middle East, and parts of Asia in which major river basins no longer maintain flows adequate to meet their population's thirst. These areas are experiencing a situation called *water stress,* which is any imbalance between water use and water supply. Water stress in the poorest countries within the belt threatens human lives. Dry regions in wealthy countries suffer economically—loss of crops, unhealthy livestock, depressed recreational areas—but water scarcity seldom harms human health in wealthy countries. A striking exception occurred in 2005 when

Hurricane Katrina hit the southeastern United States and caused a severe shortage of safe drinking water. The table on pages 132–133 summarizes the current status of the world's safe water.

Some of the countries shown in the preceding table have areas of adequate water and areas with serious shortages. For example, the United States and Australia appear to have fairly safe water supplies at the moment, but a large portion of the western states and much of southeastern Australia suffer severe water stress.

Some entrepreneurs want to make water a commodity to be sold throughout the world where it is needed. Jim McWhinney, a business writer, explained in 2008 in the *Forbes* online publication Investopedia (URL: www.investopedia.com), "Like gold and oil, water is a commodity—and it happens to be rather scarce. . . . Like any other scarcity, the water shortage creates investment opportunities, and interest in water is at an all-time high." Water traders have pointed out that the purchase and sale of water might be the only way to ensure all people in the world have the opportunity to acquire the water they need. Others argue that such "water for sale" would be an impossible obstacle for the poorest thirsty countries. Conservationists worry that wildlife and aquatic ecosystems may be ignored as people compete for the globe's remaining drops of safe water. In a 2001 issue of *Backgrounder*, a journal published by the Institute for Food and Development Policy, a municipal workers' union in southern Africa countered the water-for-sale argument. The union spokesperson explained, "Water privatization is a crucial issue for public debate. Human lives depend on the equitable distribution of water resources; the public should be given a voice in deciding whether an overseas-based transnational corporation whose primary interest is profit maximization should control those critical resources. Water is a life-giving scarce resource that must remain in the hands of the community through public sector delivery. Water must not be provided for profit, but to meet needs."

COASTS AND OCEANS

Coastal pollution comes from urban, industrial, and agricultural runoff as well as from pollution at sea that washes to shore. Pollution along coasts affects shorelines, estuaries, swamps, and wetlands. Wetlands and swamps play an especially important role as natural filters of hazardous substances, preventing them from reaching the ocean. But their plants and microorganisms degrade pollutants slowly and, today, they cannot keep up with the pollution pouring into them.

The World Water Council reports on its Web site (URL: www.worldwatercouncil.org) that 50 percent of the world's wetlands have disappeared in the last century; the contiguous United States has lost 53 percent. The United States currently retains slightly less than 104 million acres (42 million ha). According to the U.S. Fish and Wildlife Service's most recent report (2005) on wetlands, the total acreage of wetlands has increased slightly in this country. But they remain threatened in many parts of the United States and around the world. Much of the threat comes from the almost 3 billion people living near coasts, 45 percent of the world's population. Coastal areas have been overwhelmed by the stresses put on them by

The *Exxon Valdez* grounded on Bligh Reef in Alaska's Prince William Sound in March 1989. One of the most environmentally damaging oil spills in history, the cleanup involved recovering oil from the damaged ship, salvaging parts of the tanker, and collecting oil from the water's surface with skimmers and booms. Workers also tracked the oil to beaches and cleaned shorelines and rocks and rescued and cleaned as much wildlife as possible. *(E. R. Gundlach)*

human populations. In 2006 *National Geographic* quoted Admiral James D. Watkins's comments to Congress on the nation's coasts: "Our failure to properly manage the human activities that affect the nation's oceans,

MAJOR SHIP ACCIDENTS WITH OIL SPILLS (1967 TO PRESENT)

SHIP	DATE	LOCATION	COAST AFFECTED	TONS (U.S.) OIL SPILLED
Atlantic Empress	1979	West Indies	Tobago	321,440
ABT Summer	1991	Angola	700 miles offshore	291,200
Castillo de Bellver	1983	South Africa	Saldanha Bay	282,240
Amoco Cadiz	1978	France	Brittany	249,760
Haven	1991	Italy	Genoa	161,280
Odyssey	1988	Canada	700 miles off Nova Scotia	147,840
Torrey Canyon	1967	United Kingdom	Cornwall	133,280
Sea Star	1972	Oman	Gulf of Oman	128,800
Urquiola	1976	Spain	La Coruña	112,000
Hawaiian Patriot	1977	United States	300 miles off Honolulu, Hawaii	106,400
Independenta	1979	Turkey	Bosphorus	106,400

Note: Reported spill sizes vary depending on the source. The data above are from the International Tanker Owners Pollution Federation (www.itopf.com). To calculate the amount of waste in metric tons, multiply by .907.

coasts, and Great Lakes is compromising their ecological integrity . . . threatening human health, and putting our future at risk."

Pollution from recreational activities is particularly severe in coastal areas and some of it harms animal and plant life that live nowhere else but in these ecosystems. The types of refuse threatening this biota include plastic bottles, plastic six-pack holders, cigarette butts, coins, sharp metals, lead fishing sinkers, fishing line, products such as suntan lotions, food waste, clothing, Styrofoam pieces, rubber and plastic toys, mechanical parts from boats, and engine oil. The U.S. Office of Technology Assessment estimates more than 1 million birds and 100,000 seals, sea lions, otters, dolphins, porpoises, whales, sharks, and turtles are killed each year by discarded plastics alone. The Marine Conservation Society's spokesperson Emma Snowden said in 2007, "Of all the hazardous materials littering our seas today, plastic poses the greatest threat." Intentional and unintentional boating accidents and shootings add another level of threat to marine mammals, and it may be a surprise to many people to learn that balloons released over the ocean and ingested by sea mammals cause starvation by blocking their digestive functions.

Oil pollution is also recognized as a major threat to aquatic ecosystems and to people's livelihoods. Oil slicks washing ashore damage tourism and fishing industries, and seasonal jobs disappear when beaches close. Eventually, they affect entire local economies. A foundered tanker, an offshore rig explosion, or a pipeline rupture is a point source of oil pollution easily recognized. Nonpoint urban runoff, however, is the main source of oil that pollutes fresh and marine waters. Nonpoint sources can be difficult to find and control. Examples of nonpoint runoff are the following: vehicle engine leaks, service stations, illegal dumping of motor oil into storm drains and sewers, parking lots, roadways and driveways, junkyards, leaks from boats in marinas, and older unlined landfills containing auto parts.

Oil comes out of the ground as crude oil, which is a rather undefined natural resource containing many impurities. Refineries remove the impurities and convert crude oil into useful forms collectively called *refined oil*. Refined oils consist of gasoline, lubricant motor oil, heating oil, kerosene, and diesel fuel. The proportion of various long-chain hydrocarbons in oil determines the final product's composition; heavy oils contain longer chains up to 50 carbons and lighter oils contain chains of eight (octane is C_8H_{18}) to 20 carbons. The oil industry produces light, medium,

(continues on page 140)

CASE STUDY: VOLUNTEERS TACKLE OIL-DAMAGED SHORELINES

In 1969 an oil platform in California's Santa Barbara Channel exploded, spilling 200,000 gallons (757,000 l) of crude oil. Dolphins suffocated due to oil-clogged blowholes; seals with damaged fur succumbed to hypothermia; shorebirds and waterfowl died from drowning or hypothermia. Interior Secretary Walter Hickel said when he arrived on the scene, "I knew it was bad, but I didn't expect anything like this." Observers soon understood a hard lesson: The environment needed hands-on protection.

Following the accident, rig workers feverishly filled the damaged shaft with a mudlike chemical, then plugged it with cement. Boats skimmed the slick from the surface and planes dropped detergents to break it into pieces. Volunteers swarmed thirty miles (48 km) of spoiled beaches, threw straw on top of the oil, and raked up the oil-soaked material. Others melted the oil that had stuck to rocks by using high-powered steam jets. Meanwhile animal care centers, including the zoo, recovered oil-coated birds and marine mammals. Volunteers bathed the birds in the detergent Polycomplex A-11 to remove the oil, then kept the survivors warm under heat lamps. Despite the best efforts, only 30 percent of the affected birds lived.

The 1969 spill cleanup was well-intentioned but not efficient. Today emergency response leaders assess the type of oil in the spill before cleanup begins. Cleanup crews often leave oil slicks intact rather than dispersing them, and drag them from the water using skimmer boats pulling booms. Booms are inflatable floating fences that surround and capture surface oil; skirts are similar but work below the surface. Refineries clean some of the recovered oil so it can be used. Any remaining slicks used to be burned but this practice is rarely followed nowadays to prevent air pollution.

On land, alternate eco-friendly techniques are taking over. For example hot-water jets once used for melting oil on shore can kill tide pool biota. These have been replaced by special adsorbent plastic sheets that pull oil from surfaces and are then collected by cleanup teams. A nonprofit organization called Matter of Trust has begun an innovative approach to oil-spill cleanup: mats made of hair, which draws oil to it. The idea's originator, Lisa Gautier, supplied hair mats to help with cleanup of an oil spill in San Francisco Bay in 2007. Ms. Gautier supplied about 1,000 doormat-sized hair mats for volunteers to use in soaking up oil from sandy beaches. Local mushroom expert Paul Stamets helped turn the mats into a truly innovative invention: he seeded each of Gautier's mats with oyster mushrooms, which grew on the mats and absorbed the oil. "You make it like a lasagna," Gautier said to the *San Francisco Chronicle*. "You layer the oily hair mats with mushrooms and straw, turn it in six weeks, and by twelve weeks you have good soil." The hair-mushroom mats have received interest from national media outlets and environmental organizations as a feasible way to clean up oil spills.

This oiled otter was one of thousands of injured animals found soon after the 1989 *Exxon Valdez* oil spill in Alaska. Pollution harms wildlife directly and also affects ecosystems such as the ecosystem that depends on the otter as both a predator and as a food source for other predators. *(Greenpeace/Merjenburgh, Greenpeace)*

Animal rescue has also improved since the Santa Barbara spill to the point where about 60 percent of birds and mammals can be saved. Because oil-injured animals are vulnerable to shock, only trained wildlife specialists capture oiled animals and deliver them to rehabilitation centers for cleaning by trained volunteers. Typical cleanup techniques for oiled birds follow a standard series of steps:

1. Deliver to triage center for weigh-in, physical exam, and tube feeding.
2. Cover with towel and allow bird to stabilize; transport to cleanup center.
3. Rinse eyes and ears; take temperature; test blood for toxins; give fluids and tube feeding.
4. Deliver to quiet, warm stabilization room for 24 to 48 hours, with feeding and watering.

(continues)

(continued)

5. Clean with warm water and diluted dishwashing liquid; rinse, pat dry.
6. Exercise and stabilize in outdoor pond or pool.
7. Release to a clean habitat in seven to 10 days.

Oil spill cleanup has advanced since 1969 but it does not reach 100 percent success. Very thin oil sheens evade skimming techniques, and heavy winds and waves reduce the effectiveness of booms. Vigorous mixing can create an air-water-oil mixture called a mousse that contains too much water to ignite, yet contains too much air to easily collect. The *Exxon Valdez* cleanup was complicated by the large volume of mousse created in the rough waters of Prince William Sound. Spills in choppy waters now receive chemicals called *surfactants* that disperse oil into tiny droplets. The droplets reach shore where microbes begin to degrade them. Gelling agents collect smaller spills in choppy waters, then cleanup crews manually pick the oil-gel semisolids out of the water or capture the material with booms and skirts. They place collected material on conveyer belts, which carry it onto the skimmer boats. There, a presslike machine squeezes the oil out of the collection material. In this way, cleanup teams recover some spilled oil to be sent to a refinery, while the rest goes to hazardous waste disposal.

(continued from page 137)
and heavy refined oils for specific uses. For example, heavy oils lubricate engine crankcases, and light oils serve as gasoline for cars.

Oil hydrocarbons are only slightly soluble in water and lie on top of a water surface as a sheen or as tiny droplets. When pollution enters an ecosystem, microbes at the oil-water interface degrade crude oils more readily than refined oils because crude oils are mostly composed of straight chain hydrocarbons that microbial enzymes naturally attack. Refined oils contain differing levels of branched chain hydrocarbons that are more resistant to microbes. The ease with which microbes digest crude oil may provide a reason for why marine animals caught in crude oil spills have recovered faster than animals exposed to refined oil.

In the ocean, heavy oils sink and cover bottom-living organisms, such as shellfish, and they kill coral reefs. Light oils floating on the sur-

Under the federal government's National Contingency Plan (full name is the National Oil and Hazardous Substances Pollution Contingency Plan), the EPA responds to oil spills in inland waters and inland ports and harbors, and the U.S. Coast Guard manages coastal waters consisting of bays, tributaries, inlets, coasts, deepwater harbors and ports, and the Great Lakes. A national response system has been refined since the *Exxon Valdez* accident. The Marine Spill Response Corporation (www.msrc.org) connects a national network of trained workers who react quickly to any spill. Staffing shortages and scarce funds may thwart some cleanups, but there is hope that with each oil spill, public and private responses will improve. In 1969 *Santa Barbara News Press* editor Thomas Storke commented, "Never in my long lifetime have I ever seen such an aroused populace at the grassroots level. This oil pollution has done something I have never seen before in Santa Barbara—it has united citizens of all political persuasions in a truly nonpartisan cause." The Santa Barbara cleanup showed the power of community effort in saving the environment.

Santa Barbara's oil spill taught lessons about how official agencies and the public should best respond to oil accidents that damage coastal waters. Trained volunteers remain vital for saving animals and beaches in most big oil spills. Fast response is also critical. The 1969 spill had another impact; it provided momentum for the fledgling environmental movement and the country's first Earth Day.

face kill tiny organisms that form the base of food chains. Surface oils also damage the fur or feather coats of mammals and birds, respectively. Animals covered in oil cannot regulate their body temperature and soon lose body heat, a condition called hypothermia. Those not dying from hypothermia often die from ingesting the oil, which poisons them. Oil pollution cleanup is receiving keen attention due to several spectacular maritime accidents (see the table on page 136), yet with all the advances in cleanup technology, most oil in coastal and ocean spills is never recovered. If oil drilling expands to the Arctic region, as oil companies seek to do, spills on ice will create further complications in cleanup. Leslie Pearson, Alaska's director of oil spill prevention and response explained in a 2008 issue of *Business Week,* that an ice spill "takes an oil spill response to another level, as opposed to dealing with an open-water scenario."

Aquatic ecosystems pay a large price to daily oil-laced runoff as well as large and sudden oil spills from tankers, rigs, refineries, and pipelines. Though runoff delivers more overall pollution to the oceans, disasters like the *Exxon Valdez* accident described in chapter 3 (which ranks 35 in the list of largest ship oil spills) highlight the damages done to invertebrates, fish, waterfowl, shorebirds, and land and marine mammals living and feeding near coastlines. Large-scale spills provide one plus: They give cleanup experts the opportunity to improve on bioremediation and bioaugmentation techniques. Cleanup crews use better methods for removing oil from sand and rocks with each new spill, and animal recovery methods also receive much needed practice. Oil spill disasters also show the world how environmental pollution can devastate an ecosystem in a matter of hours while the consequences linger 15 to 20 years, or even longer. On page 138, "Case Study: Volunteers Tackle Oil-Damaged Shorelines" explains some of the new techniques used in oil spill cleanups.

GROUNDWATER

Hazardous compounds leach from polluted land into the groundwaters, called aquifers, below them. These bodies of water flow very slowly so do not disperse pollutants and do not dilute them. Horizontal aquifers flow at less than a foot (0.3 m) per day; some flows have been measured at no more than several feet in an entire year. To make matters worse, plants that clean natural waters aboveground do not live underground, and subsurface conditions often limit bacterial growth because of low oxygen levels. For these reasons contaminants may stay in groundwaters for decades, and persistent compounds may remain forever.

The sources of groundwater contamination are more diverse than those found on the surface. Some of the main sources cited by the Groundwater Foundation, a nonprofit groundwater-quality resource based in Lincoln, Nebraska, are the following: agriculture, unlined landfills, corroded underground storage containers, hazardous waste dump sites, industrial spills, septic tanks and leachfields, corroded sewer lines, mine tailings, road deicing salts, percolated acid rains, oil and gasoline spills, and painting projects.

Two types of compounds have become major concerns in groundwater contamination and cleanup. One is MTBE, the gasoline additive discussed in the previous chapter. MTBE affects drinking water taste and odor at concentrations of 0.01 to 0.1 ppm. Scientists are not yet sure how it affects

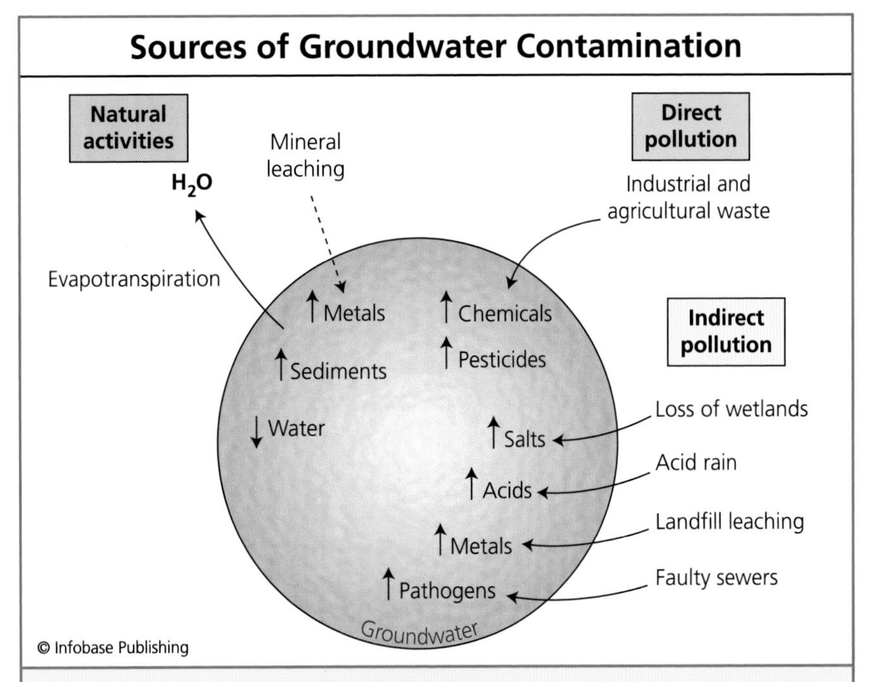

Groundwater serves as a major source of drinking water throughout the world. Pollutants that leach into groundwater damage this source, but global warming and drought also affect groundwaters. Drought causes salt intrusion; it also leads to increased metals and acids, which can ruin a drinking water source.

human health, but the EPA and individual states are taking steps to clean MTBE-contaminated waters and set MTBE limits in drinking water. MTBE makes aquifer cleanup expensive because it is very water-soluble and does not form a *precipitate* that can be collected. Absorbent carbon filters do not capture MTBE as they do other organic compounds. Current MTBE cleanup uses extraction through wells drilled into the contaminated water followed by air stripping in which contaminants are drawn from water as a vapor. Air stripping does not remove all MTBE, however, and scientists are turning to oxidation technology for remediating MTBE-contaminated water.

The second type of problem contaminant is the group known as dense nonaqueous phase liquids (DNAPLs). DNAPLs are chlorinated solvents used since the early 1900s in industrial processes. They serve a variety of purposes: metal and plastic cleaning, paint stripping, degreasing, dry cleaning, and extraction chemistry; as chemical components of fire extinguishers and adhesives and aerosols; and in pharmaceutical manufacture and foams and fluorocarbon production. DNAPLs are heavier than water and contaminate underground sites, they are persistent once they enter the environment, they form strong attractions with soil particles, and they are usually found in mixtures with other hazardous compounds. In situ thermal wells remove DNAPLs from soils. These thermal wells raise soil temperatures to 750°F–1,300°F (400°C–700°C) and drive the compounds out as vapors. An outflow well then collects the vapors and condenses them into a form that can be treated by incineration.

As mentioned, Superfund and brownfield projects often need the fastest cleanup methods possible. For groundwaters, that means pump-and-treat, in which the contaminated water is pumped out of the ground, put into tanks, and shipped elsewhere for removal of hazardous chemicals. The University of Pittsburgh has developed guidelines for cleanup technologies that will be more efficient than pump-and-treat, such as the following: in situ air sparging, vapor stripping, thermal wells, horizontal wells, in situ flushing, chemical stabilization/solidification, surfactant treatment, ex situ ultraviolet treatment, ex situ oxidization, electrokinetcs, and biological remediation. In addition, microbes that grow well in sediments can be developed in laboratories for groundwater bioremediation. For instance, bioslurping is a biological cleanup method in which technicians flush air down a well drilled into the contamination. The air then boosts the growth of native soil bacteria—an intrinsic mode of bioremediation—and the bacteria then degrade the contaminants.

SURFACE WATER TREATMENT

The Clean Water Act of 1972 sets a goal of zero discharge for industries and municipalities, meaning they must assure that they put no contaminants directly into lakes, ponds, streams, or rivers. For zero discharge to work, towns must keep their sewer systems maintained and free of leaks. Wastewater treatment plants must also assure that treated wastewaters carry no hazards back into the environment. But no one can control storms. In heavy rains, runoff can overflow the storm sewers that lead to treatment plants. In the worst cases contaminated waters bypass treatment altogether and enter rivers, estuaries, and coastal waters. Storm surges from the ocean further compound the problem by pushing contaminants back upstream and flooding coastal areas. There are, however, technical and natural means for minimizing threats to clean water sources.

Sustainable water use is a concept that combines water remediation technology with waste prevention. It includes ways to help the land minimize the harm caused to water sources by storms and other natural events. Sustainable water use also involves the reduction of water waste by every individual and business. Advanced techniques then remove contaminants and restore water for humans, plants, and wildlife. Lastly, sustainable water use requires that wetlands be conserved so they can act as sponge-like buffer zones to protect against flooding and filter out organic compounds.

Desalination is a process in sustainable water production, which removes salt from marine waters and brackish (low in salt) waters. Desalination plants use either of two methods to reduce salts in marine water (about 35,000 ppm dissolved materials) and brackish water (1,500–6,000 ppm) to potable (drinkable) levels of 500 ppm or less. The first method is distillation, a simple procedure whereby freshwater evaporates from salty water when heated. Water treatment plants then remove impurities from the cooled and condensed water so it is suitable for drinking and bathing. The second method is reverse osmosis. In this method, pumps send salt water through large filters under high pressure. These filters, called semipermeable membrane filters, hold the salts and allow the desalted water to flow through. It is called reverse osmosis because salts collect on the side of the membrane containing a high salt concentration. In regular osmosis, like that found in living cells, salts move in the direction of lower salt concentrations.

Desalination includes some disadvantages that engineers must find ways to overcome. For instance, desalination plants sometimes drain brackish waters that contain sensitive ecosystems. The water discharged from desalination plants also carries a high level of salt and may harm habitats where it is released. It causes a particular threat to ecosystems on the seafloor because dense salt water sinks below less-concentrated water, such as marine water. Finally, desalination plants demand high amounts of energy.

The first desalination plant was built by the British in 1869 near the Red Sea to replenish the water needs of the Royal Navy. Today, Saudi Arabia, the United Arab Emirates, Kuwait, Spain, and Australia collectively operate thousands of desalination plants. The United States has experimented with desalination as a backup to traditional water treatment, but no U.S. community currently relies entirely on desalination to supply its drinking water. California, Texas, and Florida have led the way in developing desalination programs to supplement their municipal drinking water.

In 2008 the *Tampa Tribune* reported that Hernando County, Florida, had about a ten-year supply remaining in its groundwater sources. *Tribune* reporter Tony Merrero wrote of the situation, "With a burgeoning population and a limited groundwater supply, the authority needs the extra money to work with its member governments to come up with alternative water supply projects for this region such as reservoirs and desalination plants." Indeed, the costs of desalination equipment and the plant's energy needs have been major reasons why few new desalination plants have been built in the United States. Perth, Australia, has tackled the high costs of desalination—the city is located in one of the world's most water-stressed areas—by using wind farms to power the desalination process. In the June 18, 2007, National Public Radio show called *Climate Connections,* the plant's manager, Kerry Roberts, said, "If you look at the combined output of the wind farm at maximum wind speeds—24 to 28 miles per hour (24–48 km/hr)—you're looking at an output of close to 80 megawatts." Roberts pointed out that the output is enough power to run the city's desalination plant 160 miles (257 km) away.

THE GLOBAL OCEAN

Every aspect of civilization has been related in some way to waters inland and in the oceans. Globalization has increased the amount of trade taking place over the open seas and spread industries into places that previously lived agrarian lifestyles. Agriculture also affects faraway places. In 1996 a Brazilian team of ocean chemists demonstrated how the condition of water in one place affects ecosystems halfway around the world. They showed that penguins had accumulated pesticides carried in their prey. These chemicals had probably drifted thousands of miles by sea to Antarctica where they ended up in the birds' fatty tissues. (Pollutants carried by wind can also contaminate the ocean in distant places when the pollutants drop out of the air.)

Geographers identify five oceans on Earth: Atlantic, Pacific, Indian, Arctic, and Southern. In ecology they are actually all one body interacting with the atmosphere to shape local climates, which influence local ecosystems. The oceans undergo natural warming and cooling cycles that bring drought to some regions and cause floods in others. Ocean conditions normally change over many years or several decades. This resistance to sudden change is called *momentum inertia,* and probably saves ecosystems by allowing time for them to adapt to natural warming and cooling cycles. But momentum inertia may intensify the results of pollution because ocean ecosystems killed by pollution do not return within a single season.

Coral reefs provide an example of momentum inertia. Coral reefs occupy 0.1 percent of the ocean, yet these ecosystems support at least 25 percent of all marine species, almost 70 percent of marine fish alone. At least 50 different types of organisms representing more than 1,000 species act as producers, consumers, or decomposers within reefs. Coral reefs further protect a large portion of the world's coastlines by absorbing energy from storms, and they prevent global warming by removing carbon dioxide from the air. But 20 percent of the Earth's coral reefs have been permanently ruined; over 70 percent are threatened. *Time* magazine's online publication of August 17, 2007, interviewed Gregor Hodgson,

Major Roles of the Global Ocean

Role	Effects on Biota
currents	wind, climate, local weather, water temperatures
water cycle	supports all biological life
heat reservoir	energy store, climate and weather
carbon reservoir	supports biochemical reactions, removes carbon dioxide
tides	energy, food source, habitat replenishment
ice caps	water storage, atmospheric temperatures
biodiversity	about 250,000 known species
resources	mineral deposits, nutrients, oxygen

the executive director of the Reef Check Foundation. Said Hodgson, "I grew up diving and snorkeling all over the world. Those reefs are all gone." Almost all of the damage has been from human activities: coral mining, fishing, oil spills, agricultural runoff, and industrial pollution. In Southeast Asia fishermen blast reefs with explosives to stun fish and make them easier to catch, or they inject cyanide to flush species out of the coral and into their nets. Large trawlers scoop tons of bycatch, or unintentionally caught fish, that contains varieties essential to reef ecosystems. Lastly, warmer ocean temperatures have killed corals and the dead reefs turn a chalky white in a phenomenon known as *bleaching.* Corals can adapt to change, but they apparently cannot adapt fast enough to avoid these many threats. Hodgson has tried various tactics to get coral reefs on the World Conservation Union's endangered species list, pointing out that several anticancer drugs come from reef species. "Maybe one day a coral will save your life," he said. "That gets to people." Rescue of the coral reefs presents an opportunity for countries to cooperate in restoring the global ocean. Every place on Earth has a stake in it, as shown in the table on page 147.

Rachel Carson wrote the following eloquent tribute to the global ocean in her 1961 book on ocean ecology, *The Sea around Us*:

> For all at last return to the sea—to Oceanus, the ocean river, like the ever-flowing stream of time, the beginning and the end.

CLEAN WATER TECHNOLOGY AND LAW

Thousands of organizations and universities focus on the status of the world's waters and they follow three main objectives: exploration, conservation, or pollution. All of them agree on one purpose of water quality programs: to assure the planet's biota have a permanent and dependable source of safe water to sustain life. This task relies on new technologies.

It is often said that scientists have learned more about outer space than they have about the depths of the ocean. Technologies now used for ocean studies have been adapted from space exploration, and vice versa, and oceanographers have begun to apply to aquatic ecosystems the methods used for repairing land ecosystems. Jeremy Jackson is a world-renowned marine ecologist at the Scripps Institute of Oceanography in San Diego, California. In a lecture at Harvard University in 2008 titled "Brave New

Ocean," he summarized the state of the world's waters as follows: "The oceans are no longer wild and no place is pristine. Six major changes harbinger the future of the oceans: (1) loss of big animals, (2) denuding of the ocean floor, (3) globalization of biota due to extinctions and introductions, (4) ocean warming, (5) increased pollution of all kinds, and (6) . . . the so-called 'dead zones' proliferating in coastal seas around the world."

Water technology methods fall into three categories based on their objectives. First, surveying technology consists of the satellite images and samples collected by submersible vessels. Second, analytical technology encompasses the techniques discussed in chapter 1. These are used for defining water conditions and measuring contaminants. Third, cleanup technology consists of mechanical or manual means of removing water contaminants. As examples, a boom is a mechanical tool for removing oil from water, but bathing oiled animals represents a manual method for cleaning up pollution.

International organizations provide protections to the global ocean in two ways. First, they promote the development of new technologies such as satellites and underwater vessels. Second, they encourage cooperation between countries for protecting Earth's waters. Each organization has different objectives in water protection. The Organisation for Economic Co-operation and Development, for example, follows the impact of water quality on national economies. The WHO tracks the effects of poor water quality and scarcity on human health. The United Nations promotes the well-being of each nation's citizens and a safe clean source of water helps meet this objective. The World Wildlife Fund has made water quality a priority for maintaining ecosystems and preserving biodiversity. These organizations do not enact laws and they cannot force citizens to comply with national environmental laws. They do, however, create strong alliances between countries to safeguard human health and a healthy environment.

Countries have combined their talents in water technology for cleaning polluted surface waters and groundwaters. The UN's World Meteorological Organization (WMO) is an example: It provides countries with climate information as it impacts local water status and provides information on improving water sources. The WMO also leads sea studies and has expertise in the following survey technologies: hydrometry instruments for measuring compounds in water discharge; ocean drilling and deep-seafloor sampling; operating undersea laboratories; and satellite imagery. WMO Secretary-General Michel Jarraud explained the organization's contributions in a March 2008, press release: "National Meteorological and Hydrological Services can significantly reduce the impact caused by such extreme weather events by providing . . . information, including forecasts and early warnings, to governments, the public, and the media. It is not possible to prevent natural hazards, but the loss of lives and the damage that they cause can be minimized through risk management based on better observations." The European Union (EU) also supports programs for finding new technologies. In 2000 the EU established the Water Framework Directive, which sets out a plan for countries to follow when analyzing water and managing water protection programs. The Directive covers restoration plans for inland waterways and lakes, groundwaters, coastal waters, and transitional waters (low-saline waters at river mouths).

In the United States, the National Aeronautics and Space Administration (NASA) uses satellites to measure tides, temperatures, salt circulation, algal blooms, dissolved organic compounds, and phytoplankton levels in the ocean and in estuaries. Color scanners aboard each satellite detect the wavelength of light emitted from the water's surface. For example, scanners measure phytoplankton by sensing the wavelengths absorbed by the pigments chlorophyll *a* and phaeophyton *a*. Computer algorithms convert color readings from infrared wavelength detection to temperature ranges. Organizations such as the Woods Hole Oceanographic Institution and the Monterey Bay Aquarium Research Institute combine satellite data with data collection buoys fitted with sensors to detect and map the agricultural pollutants nitrate, ammonium, and phosphates. Perhaps the most advanced undersea methods of all have been carried out by the underwater laboratories of the National Oceanic and Atmospheric Administration (NOAA) that hold up to six aquanauts. Aquanauts are scientists that live aboard submersible vessels while they conduct studies on ocean conditions. Typical vessels work at 85 feet (26 m) underwater and withstand pressures two and a half times that at the surface. The submerged research station, about the size of a railroad freight car, allows aquanauts to e-mail study data by a signal that reaches a buoy at the surface. The buoy relays the information to laboratories on land. *National Geographic* reporter Gregory Stone wrote of his experience in such a vessel named *Aquarius*

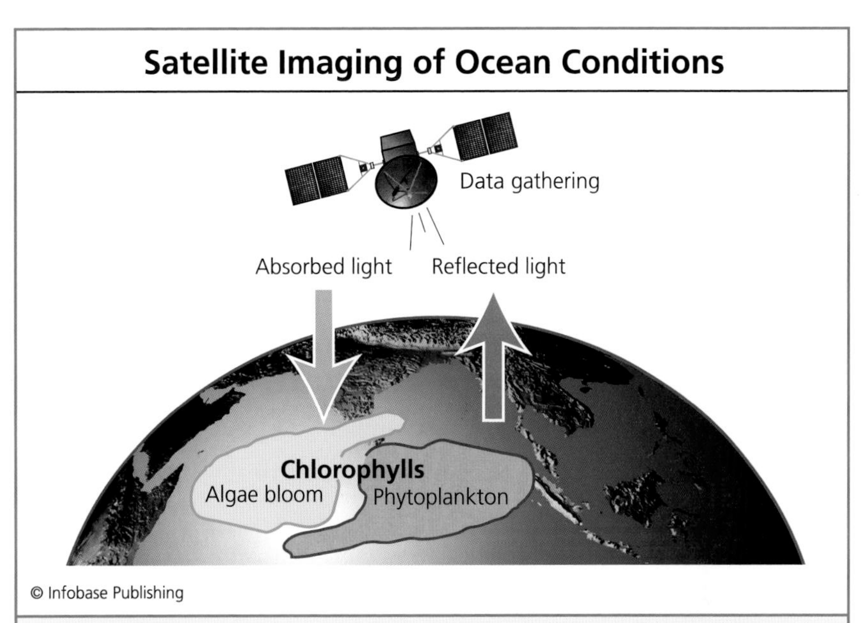

Satellites used in environmental studies carry equipment to monitor ocean ecosystems. The major data collected by satellites pertains to phytoplankton levels, water temperature, organic nitrogen compounds, inorganic metal compounds, and algae blooms. Global positioning systems enable scientists to create accurate maps from the satellite data.

in 2003, "Living in *Aquarius* was like a spaceflight, a submarine ride, and a week in a college dorm wrapped in one." Scientists working from submerged vessels can put tracking tags on fish, monitor the patterns of fish living in coral reefs, conduct *in vivo* experiments on reef samples, and make observations on the behavior of other marine creatures.

Chapter 1 explained how new environmental laws could prompt scientists to find better analytical methods. Advances in water technology likewise lead to better environmental laws. In the United States, the Safe Drinking Water Act (SDWA) of 1974 has led to better methods for monitoring aboveground and underground water. The American Water Works Association (AWWA) helps local utilities adapt the best new water technologies for meeting stricter SDWA requirements. Internationally, many countries follow the Bonn Charter for Safe Drinking Water, known as the Bonn Principles. These Principles were set by international water quality experts to address all areas of drinking water treatment and distribution. The key Bonn Principles are the following:

- Water management should involve consideration of the entire water cycle.
- Methods to ensure safe drinking water should assess quality all along the distribution system and not solely at the end of the system (the tap).
- Water quality systems require cooperation and communication among government, regulatory authorities, environmental groups, contractors, plumbers, and consumers.
- Water should be safe, reliable, and aesthetically acceptable.
- The price of water should be set so that it does not prevent people from having adequate water quantity and quality.
- Water quality practices should use the best available scientific evidence and technology.

In the United States the Marine Protection, Research, and Sanctuaries Act (MPRSA) of 1972 came about in response to the increasing incidence of oil tanker spills as the United States searched the globe for new oil reserves. Each spill added to the already alarming levels of industrial wastes entering the ocean. At that time companies around the world poured chemical and low-level radioactive wastes into the sea, and others added ash from at-sea incineration of solvents and halogenated compounds, mine tailings, scuttled ships, large steel and concrete structures, sewage sludge, and dredging sediments. The MPRSA now ensures the protection of shorelines, coastal waters, and offshore waters. Nicknamed the Ocean Dumping Ban Act, this law bans ocean disposal of industrial wastes and the sludge produced by wastewater treatment plants. Today the only materials the United States permits to be discarded in the ocean are sediments dredged from ports or navigation channels, but even this form of disposal may cause some harm because dredged materials deliver pollutants to nonpolluted areas and might interfere with the normal activities of bottom-dwelling marine life.

Another important step took place in 1972. In that year members of several countries met in London to develop new rules for protecting the oceans through international laws in response to the alarming recurrences of oil spills, radioactive dumping, scuttled ships, and billowing clouds from offshore incinerators. The meeting became known as the London Convention, and eventually grew to 80 members, including the United States. The UN's International Maritime Organization (IMO) has implemented the convention's rules since 1975; these rules focus mainly on the protection of coastlines from oil spills. Most of the IMO's requirements focus on oil tankers, as shown in the following list.

- Empty tanks are filled with inert gas to prevent explosions.
- Each ship carries backup fire-extinguisher and navigation systems.
- Ship design strives for increased stability after a collision or other accident.
- Each ship is equipped with hitches for emergency towing.
- Ships are built with double hulls to reduce the chance of leaks during accidents.

About 10 years after the London Convention, the United Nations called together the Convention on the Law of the Sea. The Law of the Sea (LOS) became a set of regulations for managing ocean navigation rights and fishing rights. Since the 17th century coastal nations have claimed and conflicted over the waters off their shores, first over fish stocks and right of passage, and then over mineral resources and valuables lodged in shipwrecks. Today the LOS sets territorial limits and related maritime rules followed by more than 150 countries. The LOS works in conjunction with the other laws mentioned here to protect the Earth's surface waters, yet much more must be accomplished to prevent the destruction of the world's water supply.

CONCLUSION

Surface waters and groundwaters throughout the world have been contaminated with hazardous wastes. This contamination can be chemical, physical, or biological in nature. Many technologies are available to remediate polluted surface and underground waters, but the urgency to find more effective techniques is increasing. Climatic changes and growing populations have put severe stress on the world's water availability. Today, many cultures have less water than they need to maintain heath, a situation called water stress.

Water pollution comes from point and nonpoint sources. Waterways are integral parts of transportation, industrial needs, religion, and recreation, and they serve as drinking water sources for many communities. Groundwaters are similarly threatened by pollution and, because they are slow-moving, pollution can stay in them for a long time. Groundwaters are also more difficult and more expensive to clean than surface waters.

Oil runoff and spills create a significant environmental threat to marine life, but more oil pollution comes from urban runoff than from large and dramatic tanker spills or rig accidents. Techniques for cleaning oil-marred beaches and coastal waters are improving. Even so, none of the current methods remove all oil pollution.

Advanced technologies have become available for water cleanup and water monitoring. Prevention of water contamination is critical because contaminants can remain in oceans and in sediments for many years. Freshwater bodies are protected by national agencies, while marine waters and ecosystems are protected by international laws.

Water remediation technology is one of environmental science's priorities. Not only are ecosystems and wildlife threatened by water scarcity, but human conflicts can be expected to increase in the near future if safe water supplies become drastically threatened.

Superfund Sites

Few environmental laws are more famous than the 1980 Comprehensive Environmental Response, Compensation, and Liability Act (CERCLA), which established a large trust fund known as Superfund for pollution cleanup. Superfund's resources target the nation's worst hazardous waste sites. Unlike brownfield cleanup, which is done voluntarily by a land developer, Superfund mandates that polluters be responsible for cleaning up their environmental wastes. Failing to comply is against the law.

Controversy has reigned throughout Superfund's history. Almost any U.S. law that affects thousands of industries, millions of businesses, and billions of dollars must overcome its share of debate. Congress, industry, and environmentalists constantly question the effectiveness of Superfund. Businesses worry about cleanup costs; environmentalists question the long-term effects of a site's pollution and its cleanup on ecosystems. Politicians hope to keep both sides happy. Communities located near Superfund sites may have the most at stake due to the hazardous substances in their vicinity. Taken together it may not be surprising that over a quarter of Superfund expenditures today is spent on legal costs.

CERCLA addresses the national history of indiscriminate waste dumping and poor waste management. Superfund does not distinguish between contamination made by intentional waste dumping and contamination caused by accidents or ignorance. Superfund's original intent was to focus first on the country's biggest environmental problems, then deal with lesser contaminations. Early in Superfund's history, scientists and the federal government realized that the problem of contamination stretched much farther across the land than anyone had originally thought. No single law such as Superfund could likely fulfill the needs of the U.S. economy, national well-being, and the health of the planet. Yet Superfund is the United States's most ambitious effort to reverse decades of environmental damage and balance its intimidating challenges with its enormous promise.

This chapter describes the checkered history of Superfund, its successes, some failures, and its importance in cleaning up the nation's worst hazardous waste sites.

HISTORY OF SUPERFUND

Before toxic dumps burst into flames and smokestacks belched horrible emissions into the air, toxic wastes were carried out back on company property, dumped, and forgotten. Manufacturers dug trenches or carved depressions out of the earth for waste stockpiles. The dump sites became so prevalent, the Environmental Protection Agency (EPA) nicknamed them PPLs, for "pits, ponds, and lagoons." Many of these PPLs became the first Superfund sites.

In the 1960s environmentalists began to urge the government to take action against PPLs and other abandoned piles of hazardous wastes. President Richard Nixon responded to these concerns by signing into law the National Environmental Policy Act (NEPA). The new law represented the nation's first attempt to legally control polluters, and yet 10 years after NEPA became law, toxic chemicals continued to threaten people's health. In 1981 a sewer system filled with hexane exploded in Kentucky, the result of a company finding a convenient loophole in NEPA. The explosion was only one incident of many that highlighted the flaws in early environmental laws. Companies exploited the vagueness of NEPA in order to find strategies for discarding wastes without following the bothersome and costly new law. Polluters pumped tons of chemicals from huge pipes into rivers or dumped wastes on roadsides in the dead of night. The few courts that prosecuted polluters did so in isolated cases, so little national attention fell on industry as a whole. More often then not polluters wriggled free from responsibility. NEPA had possibly made the pollution crisis worse rather than better.

Government leaders agreed that more than NEPA would be needed to manage waste streams that had gone out of control. NEPA's biggest flaw was that state and local governments had little power to enforce it. In 1976 Congress gave more power to local governments and added more controls to waste management through the Resource Conservation and Recovery Act (RCRA). The law's authors also added a new category of waste to be managed: infectious hazardous waste.

RCRA addressed problems of hazardous wastes at the point where they were produced, but large amounts of decades-old wastes remained in every state. These abandoned waste stockpiles contained troubling unknowns, such as unidentified chemicals, undetermined quantities of substances, and random spread of chemicals into air, soil, and water. Congress developed CERCLA to bridge the gaps and plug the loopholes in the existing environmental laws. In 1980 President Jimmy Carter addressed the nation's waste dilemma by signing Superfund into law. For the federal government, Superfund broke new ground. Congress had never before enacted a law intended to hold polluters responsible for contaminating the environment. Perhaps Superfund also would avert costly court battles between families made sick from chemicals and the polluters, who had been taking no responsibility for their wastes. How well would Superfund work, considering that many of the nation's worst industrial polluters were powerful allies with the country's political leaders?

The EPA took over the task of making Superfund a workable program, and since the EPA's mandate included protection of the environment, this agency would be called upon to stand up to some of the world's largest industries. Almost overnight Superfund transformed the EPA, becoming the agency's biggest project. First, EPA scientists tried to identify any health effects on humans and animals from the toxic chemicals. They made attempts to rank chemicals in the order of their danger to the environment. EPA scientists also studied the scant data available to them on how chemicals migrate in the ground and in water. The EPA collected information on the best technologies for treating, storing, and disposing of various wastes. Next, the EPA took to the road to find the nation's worst contaminated sites. The agency needed very little time to realize the enormity of the problem. The EPA's technologies were inadequate and its staff was stretched to the limit by the new program. As soon as funding allowed, the EPA began hiring additional trained scientists and it retrained members of its staff to tackle the new responsibilities.

One knotty question worrying EPA scientists was a definition for "How clean is clean?" They realized that 100 percent removal of certain chemicals would not be necessary to eliminate a health threat, because some chemicals needed only to be reduced to trace amounts to make them safe. A second question related to how Superfund scientists were to know when a cleanup was complete. This is because each hazardous waste seemed to have its own characteristics. Through the 1980s the EPA studied all available information on each hazardous chemical and set a hazard ranking system for these chemicals. The EPA further invested thousands of hours into learning the acceptable limits of each chemical in air, water, and earth. Next, the EPA set up the National Priorities List (NPL) of all the worst contaminated sites in the country, to be revised yearly, and in need of the most urgent attention. Finally, the agency prepared emergency plans for local communities in case a hazardous-waste disaster such as explosion or fire occurred.

Within a few years after Superfund's start, the EPA and Congress knew that the cost of cleaning up the nation's worst sites would far exceed the funds originally budgeted in 1980. In 1986 Congress increased the trust fund from $1.5 billion to $8.5 billion. Initially, Superfund grew through taxes on chemical and petroleum industries because they had been considered the heaviest producers of hazardous wastes. By 1990 the fund was over $15 billion. It should come as no surprise that those industries paying Superfund taxes opposed this financial burden. Companies began building strong legal teams with the sole purpose of circumventing Superfund law, so by 1994 more than 20 percent of the EPA's Superfund expenses were spent not on cleanups but on legal costs.

Congress in 1995 devised a plan to solve Superfund's legal quagmire, but this plan also weakened the core principle of Superfund. No longer would responsible parties be the sole source of money set aside for cleanup funds; individual taxpayers now became the means of keeping Superfund's coffers filled. Even with this change, the inflow of money could not keep up with the cost of cleanup. Superfund today contains about $75 billion, which the Congressional Budget Office has calculated will fall far short of future needs. About 65 percent of the trust pays for site assessments and cleanup, while litigation costs have increased to 25 percent.

Superfund pays for an entire site cleanup if a responsible party is not identified. This explains why legal actions are such a large part of Superfund today; polluters fight to prove they are not at fault in causing contamination. In 2007 the Supreme Court made a ruling intended to unlock the stalemate. According to the unanimous ruling, polluters who volunteered to clean up a Superfund site before being forced to by law could recover part of the cleanup costs from the other polluters of that site, thus sharing the overall cost. On the other hand, if the EPA was forced to pursue a polluter, the polluter would become entirely responsible for all cleanup costs.

A few modifications to Superfund have been made along the way to further smooth the process. The EPA's Superfund Accelerated Cleanup Model (SACM) streamlines how cleanup sites are assessed and selected; it also includes a classification system for different types of contamination. For each classification, the EPA lists the best cleanup technologies with the hope that this information will help polluters save time in planning before cleanup begins.

As early as 1990 the EPA had learned that cleaning up a site and removing it from the NPL was not the smooth process it had envisioned. So in that year, it modified the NPL. For a site to be defined as a cleanup "success," construction on the site must be completed without necessarily cleaning up the actual wastes. The EPA's definition of construction includes activities such as road-building, warehouse and factory demolition, and drilling of cleanup wells or barriers. For a site to be moved onto the list of completed projects, the cleanup crew must put in place additional items such as lighting, security fencing, periodic debris removal, access roads, and monitoring. Therefore, a Superfund may be called a success even if no contaminants have been removed!

Today's Superfund is not perfect but it does provide a way for citizens to take action against hazardous wastes in their communities. It works like this. First, a citizen, community, or local government alerts the EPA to a potential hazard. EPA officials then visit the site and do a preliminary assessment by collecting samples and reviewing any information known about chemicals at the site. This period of site sampling is potentially dangerous because it disturbs unknown amounts of hidden hazards. In 1987 EPA spokesperson Rich Cahill described the situation to the *New York Times* this way: He said experts consider the testing period to be "a dangerous time for the surrounding area." For that reason the EPA often recommends residents near the site leave their homes during the day. A laboratory analyzes the samples by the methods described in chapter 1. Second, the EPA reviews all the information and issues a report that describes the pollution, its levels, and any known health risks. Most important, the report includes a proposed cleanup plan. The EPA also gives the site a grade from 1 to 100 (100 is the score for the worst pollution) in a procedure called the Hazard Ranking System (HRS). Sites with high scores are then put on the NPL. (The hazardous waste industry prefers the term *NPL site* to *Superfund site* as a more accurate descriptor.) At present, the EPA is reviewing thousands of sites in the United States for possible addition to the NPL. The fourth step involves listing each proposed site in a daily U.S. government publication called the *Federal Register,* which summarizes all rules and proposals by federal agencies and also any presidential documents. Any citizen may send the government a comment on the items in the *Register.* Therefore the fifth step is public comment on the proposed Superfund site. At a 2008 site evaluation in Philadelphia, EPA spokesman Roy Seneca noted, "The public will have a chance for input before work gets under way." His words summarized an important part of every Superfund cleanup in the United States. U.S. citizens have 60 days to express any opinions on the site's cleanup or its designation as a Superfund site. The sixth and final step before cleanup begins comes at the end of the comment period. At this time the government issues a letter called a *Record of Decision* to describe the site's condition, relay all public comments, and explain the technologies available for cleanup. This series of steps works better than perhaps any other waste cleanup program in the world.

As Superfund reached its 20th anniversary in 2000, more than 6,400 places had been identified for contamination assessment and over 750 had been completely cleaned up. Katherine Probst, director of Resources for the Future, an environmental policy research group in Washington, D.C., has pointed out that Superfund liability "does provide a very clear and very real incentive to manage hazardous substances properly. And that is really the purpose of a liability system, so in that sense it has been hugely successful." The NPL currently contains about 1,280 sites that still need work. Thirty of these sites have not yet received any action at all. The EPA and environmental groups publish lists of all the Superfund locations in the United States and the cleanup status at each site.

TODAY'S SUPERFUND SITES

Each of the country's Superfund sites has an HRS score based on the following three factors:

1. identification and characterization of the waste sources
2. identification of migration routes—ground water, surface water, soil, air

3. identification of the targets—people, animals, fish, habitats—of each route

Detailed items within each of these categories describe their effects on human food chains, drinking water, and the ecosystems exposed to the hazard. Numerical scores from groundwater, surface water, soil, and air pathways are combined in a single equation to calculate the final score. Any site scoring 28.50 or greater is eligible for the NPL.

New Jersey has the highest number of sites on the NPL, as seen in the table below, followed by Pennsylvania, California, and New York. North Dakota is the only state with no sites on the NPL. The total numbers of sites per state may not tell the entire story. Some states with a small number of sites could hold many tons of materials that have yet to be removed, whereas other states with a large number of sites that have been in cleanup projects for years may have less total waste remaining.

Sites on the National Priorities List (2010)

Number of Sites	States
more than 50	New Jersey (112), Pennsylvania (95), California (94), New York (86), Michigan (66), and Florida (52)
35–50	Texas, Washington, Illinois, Wisconsin, and North Carolina
20–34	Ohio, Indiana, Massachusetts, Virginia, Missouri, South Carolina, Minnesota, and New Hampshire
6–19	Colorado, Maryland, Utah, Georgia, Montana, Connecticut, Delaware, Kentucky, Alabama, New Mexico, Nebraska, Tennessee, Oregon, Rhode Island, Maine, Iowa, Kansas, Vermont, Louisiana, West Virginia, Arkansas, Arizona, Oklahoma, Idaho, and Alaska
5–0	Mississippi, Hawaii, South Dakota, Wyoming, Nevada, District of Columbia, and North Dakota

Despite many Superfund success stories, about 11 million people continue to live within one mile (1.6 km) of a federal Superfund site. (Federal Superfund sites are either military bases or weapons production facilities.) One in four Americans are within three miles (4.8 km) of these sites. People living as neighbors to hazardous sites play a critical role in Superfund's success. Communities concerned over hazardous wastes in their surroundings can contact their district representatives. State representatives then carry the message to Washington, D.C., to encourage additional Superfund spending in their district. Even though Superfund operates in a bureaucracy, which has potential for inefficiency, it slowly makes progress. Katherine Probst testified in front of the U.S. Senate's subcommittee on Superfund in 1995, describing its successes and failures. She noted, "Perhaps the one criticism on which everyone agrees is that site cleanups take too long."

Perhaps the best way to understand Superfund is to examine three sites, each at three different stages of cleanup. One project has been completed and the site is a model in restoration—a Superfund success story is one in which the original hazardous waste site has been restored to health and removed from the NPL (though it remains on the public list with a date in the column titled "deletion"). The second is undergoing the cleanup process, and the third has yet to begin cleanup, its hazardous wastes untended for years.

February 20, 2002, was a happy day for Leadville, Colorado, a small mountain town 100 miles (161 km) west of Denver. That was the day it had officially removed an environmental nightmare. Leadville's name offers a clue to its history. Since the 1859 Colorado gold rush, the 16-square-mile (28 km^2) California Gulch area surrounding the town was the country's richest source of lead, gold, copper, zinc, and manganese. Mining and smelting had dominated the local economy until 1999 when the last mine closed. Unemployment in Leadville had reached 40 percent, and meanwhile, metal-contaminated land and acid-polluted waters remained with each mine closure.

In 1994 a mining conglomerate led by the Asarco Company accepted responsibility for removing the mine wastes. At the time of the agreement in 1993 the EPA's lead attorney, Nancy Mangone, told the *Denver Post,* "I think it's a positive step. . . . It gives me hope that we will be able to reach agreements with them over the actual cleanup." Looking for ways to redeem its image in the community, Arasco worked with community leaders to poll residents for ideas on how to remediate the mining site. According to project consultant and Leadville resident Mike Conlin, the town's overwhelming response was for "a multipurpose trail that would highlight and showcase Leadville's mining history." Rather than ignore Leadville's mining past and its environmental harm, a restoration team sought a way to keep the town's connection to mining. Mayor Pete Moore was quoted in the *High Country News* as saying, "We've always seen ourselves as a mining town, and now that our last mine has closed, we're facing an identity crisis." The new Mineral Belt Trail solved this dilemma by providing trailheads that led to historic mine sites, miners' cabins, old railroads, and education centers. California Gulch today represents a Superfund project in which an ailing community revived itself. Leadville enjoys a robust new identity and it is a place where the environment is appreciated rather than damaged.

Other places are still deep in the cleanup process and hope their results will be as positive as Leadville's. Fort Ord, near California's Monterey peninsula, balances the delicate task of cleaning one section while developing

Sometimes severely polluted land can be returned to nature, such as the Mineral Belt Trail and the Rocky Mountain Arsenal National Wildlife Refuge (shown here), both in Colorado. *(Rocky Mountain Arsenal)*

other sections for profitable use. Fort Ord opened in 1917 as an army training site and remained in that role until 1975, when the Seventh Infantry Division moved in. In 1990 the EPA discovered fire retardants, paints and solvents, and poorly contained garbage dumps. The groundwaters contained VOCs and artillery ranges hid unknown amounts of unexploded ordnance, detonation devices, and lead from munitions. The EPA placed Fort Ord on the NPL that year and the base closed all its operations in 1994 during a time when the federal government decommissioned many military bases. During its site assessment EPA scientists found 11 plant communities, nine avian, and two mammalian species native to the local habitat as well as reptiles, amphibians, and invertebrates. A population of at least 100,000 people lived just outside the base and depended on the aquifer under the base for their drinking water.

Soon after the base closed, California formed the civilian Fort Ord Reuse Authority to team with the EPA, and the site's responsible party, the Seventh Infantry, to begin cleanup and restoration. With good communications between all parties, the landfills have been cleaned up and munitions are being removed from several thousand acres. Fort Ord is now divided into sectors according to the threat contained there: (1) safe and no remedial action needed; (2) restored because munitions have been removed; and (3) hazardous due to dangerous substances remaining. Fort Ord has become home to five different university campuses, two housing developments, recreational areas, and thousands of acres preserved by the Bureau of Land Management. Sensitive habitat has been restored through controlled burns and site landscaping. Several endangered lizards and amphibians such as the California tiger salamander are fighting to recover. The habitat for the endangered Smith's Blue Butterfly almost disappeared during the fort's military years, but today an area called Reserve #10 provides protected habitat for the butterfly. The remaking of Fort Ord shows that a hazardous site can be turned into both economically useful land and wildlife preserve.

The Bay City Middlegrounds, 40 acres (0.16 km^2) of land on an island in Michigan's Saginaw River, is waiting to be assigned to the NPL. Until then, a hazardous mixture of chemicals, pesticides, and metals wait in a local landfill. The landfill had for many years accepted Bay City's municipal, construction, and industrial wastes. Municipal crews then used contaminated sediments dredged from the river as

The decommissioned U.S. Army base at Fort Ord, California, has used a combination of cleanup plans. Parts of the base have been turned into classrooms for local colleges; other parts have been designated a refuge for endangered species native to the area. The remainder of the base continues in cleanup operations. *(AGSC)*

cover for the landfill, adding to the concentration of hazards in the landfill. EPA testing identified benzenes, toluene, xylenes, phthalates, pesticides, and heavy metals in leachates and waters connected to the river, which Bay City uses for drinking water. Meanwhile, fishermen work Saginaw Bay and wetlands situated in the path of surface runoff. Because Bay City purchased the Middlegrounds in 1950 and conducted the dredging and landfilling, the EPA named the town as the party responsible for its cleanup. However, the Bay City Middlegrounds has not yet been added to the NPL. Unless it is listed, Bay City will not receive Superfund money and the hazards may remain at the site for a very long time.

Some Superfund sites contain hazards even more confounding than those found in places like Bay City. On page 166, "Case Study: The Berkeley Pit" describes one such unique hazardous waste site.

CASE STUDY: THE BERKELEY PIT

Two hundred miles (322 km) from Yellowstone National Park sits a peculiar tourist attraction: the Berkeley Pit, in Butte, Montana. Mine shafts dotting Butte's landscape remind residents of the copper mining that began in the 1870s and made a fortune for a select few. The Anaconda Copper Mining Company ran all of Butte's mines for 100 years. Anaconda built at least 50 mine shafts leading to 5,600 miles (9,012 km) of horizontal tunnels, which lowered Butte's water table and required continual pumping to remove groundwaters that seeped in. The company built the Berkeley Pit in 1955 as part of its strip and shaft mining operations. In 1977 Anaconda turned the mines over to the Atlantic Richfield Company (ARCO), which extended the pit deeper into the ground as long as the pumps kept running to keep out the copper-saturated water. Copper sulfate in the pumped water was so concentrated that it supplied a good income for a nearby copper-recovery plant. ARCO's pumps chugged every day until all the mines closed in April 1982.

The town's residents probably wondered how that big hole in the ground—more than one mile (1.61 km) deep—would ever be filled and used again. Their questions soon became moot. After the last pump had gone silent, the water table began to rise. By the end of the year it had risen 1,300 feet (396 m) and metal-laced waters seeped into the pit. Water rose another 1,800 feet (549 m) by 2006. Today the Berkeley Pit contains almost 40 billion gallons (151 billion l) of acidic water filled with arsenic, copper, cadmium, cobalt, iron, manganese, zinc, and sulfates; the surface pH measures 2.5. In the cold waters below, which never go above 40°F (4.5°C), metal concentrations increase. For example, deep-water copper concentration is 190 mg/l (or ppm) and that of zinc is 620 mg/l.

Simple physics explains Berkeley Pit's problem. Groundwaters that had flooded the thousands of miles of tunnels constantly equilibrate with water levels in the pit. By 1994, 7.2 million gallons (27 million l) entered the pit from the tunnels each day—5,000 gallons (18,921 l) per minute. Water in the Berkeley Pit has been rising ever since. Today Butte's residents call it Berkeley Lake!

The EPA has responded to their understandable alarm and set a maximum level for the pit's waters. The maximum allowed level is 5,410 feet (1,649 m) above sea level. On the first day of 2007 the water level was 200 feet (61 m) below the overflow point. (The EPA and engineers have assured the Butte community that the water will never overflow Berkeley Pit because evaporation and percolation will constantly provide more space for the pit's rising volume.)

The Berkeley Pit joined the NPL in 1982: Its official name is the Silver Bow Creek/Butte Area Superfund site. The EPA eventually realized the pit was too massive to clean up completely, so it fenced in the area and ordered the cleanup work to stop. Soil near the pit is still being cleaned and Montana and ARCO monitor water levels. A battery of sensors and computers calculate the

The Berkeley Pit (as seen in 1984) in Butte, Montana, offers an example of a Superfund site that will likely never be cleaned up. Some hazardous waste sites risk more dangers in cleaning them up than in leaving them undisturbed and secure. *(Berkeley Pit, Butte, Montana)*

ever-changing flow rates into the pit due to rainfall runoff and natural groundwaters. Scientists believe the lake will reach a critical level within 10 to 15 years, and at that point, they say, it may indeed overflow its boundaries.

The Berkeley Pit Viewing Stand currently attracts a curious smattering of visitors who gaze across the greenish brown expanse and wonder . . . What do they wonder? At least one visitor, chemist Andrea Stierle from the University of Montana, wondered if life could exist in Berkeley Lake. In the 1990s she recovered live *Euglena* protozoa from the caustic water. Bacteria, algae, and fungi were soon found thriving there as well. In 2006 Stierle discovered in the pit a *Penicillium* fungus that is active against human ovarian cancer cells. In a *New York Times* interview in 2007, Stierle said, "I love the idea of looking at toxic waste and finding something of value." While studies on this potential new drug continue, bioremediation experts also look to Berkeley Lake as a valuable source of unique microorganisms that may devour toxic chemicals at other cleanup sites.

MANAGING HAZARDOUS WASTE

Cleaning up a Superfund site is a waste management project. Today hazardous waste managers need knowledge of both legal aspects and technical advances for removing contaminants from the environment. Teams of specialists also stay abreast of new cleanup technologies and the latest regulations.

Waste management includes careful tracking of the types and amounts of hazardous wastes released from a site to their final disposal in a process called *cradle-to-grave management*. In cradle-to-grave waste management, details of industrial wastes and their waste streams must by law be reported to the EPA. The EPA then summarizes the information in its listing called the Toxics Release Inventory, described in the sidebar on page 171. Some industries produce much more toxic waste than other industries, as shown in the table on page 169. The rankings of all industries

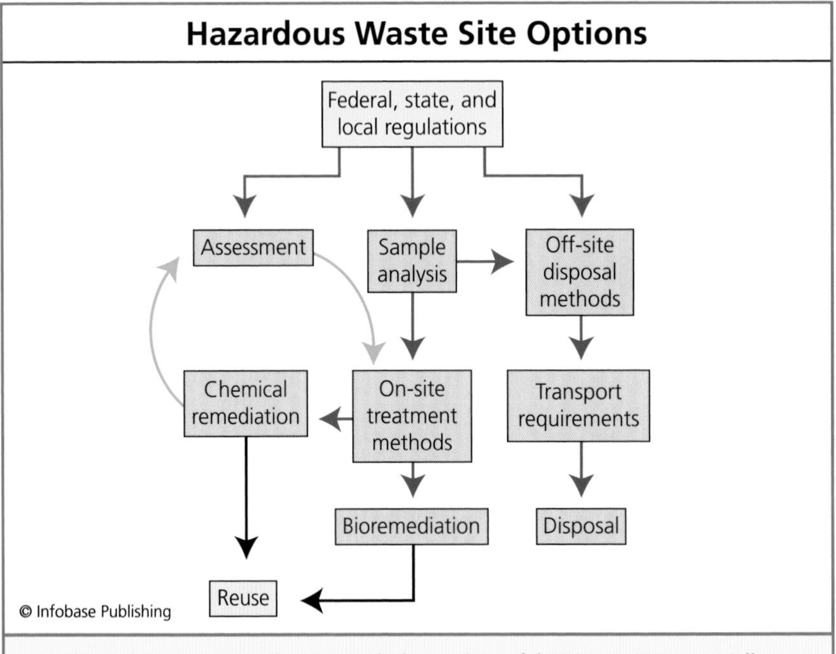

Most hazardous waste site cleanups include a variety of decision points, especially when a treatment is only partially successful. Sometimes soils may require a repeat cleanup step. Other options are the use of chemical treatment, or bioremediation, or hauling away the materials for disposal in a landfill or an incinerator.

may furthermore change as firms move manufacturing to other countries, a practice called *globalization.*

Several organizations monitor how well the government and industries do in preventing hazardous wastes from entering the environment. They provide the public with vast amounts of information on hazardous

Top Industries in Toxic Release Quantities

Industry	Total Amount of Released Hazardous Waste, Tons in Thousands (metric tons in thousands)
metal mining	580 (526)
electric utilities	452 (410)
chemicals	235 (213)
primary metals (metal raw materials)	220 (200)
paper	93 (84)
hazardous waste/ solvent recovery	84 (76)
food/beverage/ tobacco	83 (75)
petroleum	39 (35)
fabricated metals (metal products)	27 (24)
plastics and rubber	25 (23)
transportation equipment/vehicles	19 (17)

Toxics Release Inventory by Industry

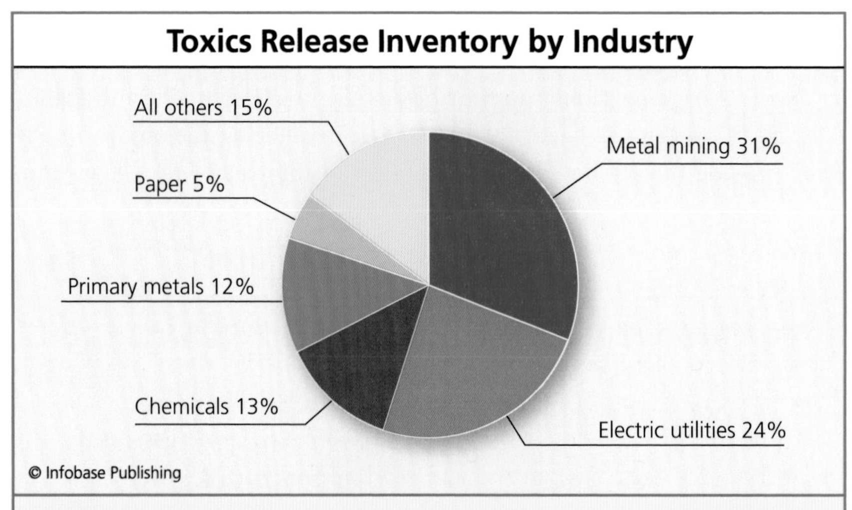

The Toxics Release Inventory provides useful information on industries most likely to release hazardous substances into the environment. The inventory's two main flaws are the following: It does not account for on-site releases and it does not include potential health risks. *(modeled after EPA)*

substances found in the TRI, and some Web sites allow users to enter zip codes to learn about releases in their vicinity. People in the United States have several resources in addition to the TRI for information on the hazards in their environment, listed in appendix D.

Waste tracking and data collection allow waste managers to see how industries perform in reducing waste releases. For instance, since the TRI has been in effect, industrial waste releases (on company property plus waste sent to other sites) have decreased by about 43 percent. In the years between 2001 and 2005 the largest producers of toxic substances (over 80 million pounds [36 million kg] yearly) decreased their total released waste quantity by 80 percent. Unfortunately, yearly amounts of wastes produced by small businesses have increased, so the EPA clearly has work to do in enforcing waste management in those places.

SUPERFUND SITE GROUNDWATER

Superfund sites have contributed to the contamination of many groundwater sources. All groundwaters are fed by surface waters and precipitation that percolate downward as little as three inches (about 7.6 cm) to as much as 900 feet (274 m). Water fills crevices, fractures, and pores in soils and rock. This hidden water serves as a source of much of the world's freshwater. As it trickles downward, the water pulls metals, organic solvents, pesticides, and caustic compounds into aquifers. Though a neighborhood in

The Toxics Release Inventory

The Emergency Planning and Community Right-to-Know Act of 1986 was written to ensure that people would be aware of the contaminants in their environment. Known as simply Right-to-Know, this act makes the EPA responsible for maintaining a list of all of the industrial toxic substances and their amounts released yearly into the environment. The list is called the Toxics Release Inventory (TRI). The inventory list is available to the public and is organized by year, state, manufacturer, chemical released, and the medium in which the substance was released (air, surface waters, burial, land treatment, etc.). In 1990 the Pollution Prevention Act lowered the minimum levels of chemicals to be reported in the TRI and also provided more details on recycling and waste reduction. The TRI is found on the EPA's Web site (URL: epa.gov/tri/index.htm).

One disadvantage of the TRI relates to the absence of small but constant waste streams. Few people think about the chemicals released in the making of a new pair of shoes, a textbook, or stack of neatly folded shirts from the dry cleaner. Wastes from immense manufacturing plants are easy to see and fairly easy to track, but small releases are difficult to track and they are not likely to be part of the TRI. Many chemicals enter the environment each year in hundreds, perhaps many thousands, of small releases that are never measured.

Laboratories invent many new chemicals each year and the EPA tries to keep up with the expanding list. In 1998 the EPA added more industries to the TRI to help monitor as many waste streams as possible: metal mining, coal mining, electric utilities, chemical wholesalers, bulk petroleum transfer terminals, and solvent recovery operations. The TRI still does not include smaller manufacturers with fewer than 10 employees or those that use less than 10,000 pounds (4,536 kg) of chemicals each year. However, the EPA in 2001 included small nonmanufacturing businesses that release hazardous wastes. Example are metal plating shops, dry cleaners, paint shops, and service stations. Even with its few flaws, the TRI makes a helpful attempt to track the diverse chemicals entering the environment each year.

the United States may be located far from any Superfund site, the aquifer supplying part or all of its water may carry pollutants from many miles away.

The Midwest's Ogallala Aquifer gives an example of a crucial and immense water source. Also known as the High Plains Aquifer, this underground body of water stretches 170,000 square miles (440,300 km^2) beneath eight states: Nebraska, South Dakota, Wyoming, Kansas, Colorado, Oklahoma, New Mexico, and Texas. It supplies drinking water for 82 percent of the population within its boundaries and irrigation for over 10 million acres (40,470 km^2) of high plains farmland—30 percent of the nation's total irrigation water for crops. A geologist made an alarming discovery a few years ago by laying a map of the nation's Superfund sites on top of a map of the aquifer: No fewer than 20 Superfund sites sat atop the aquifer.

The U.S. Geological Survey (USGS) has calculated that about half of the nation's drinking water comes from groundwaters like the Ogallala. The nation's many Superfund sites threaten a good number of these water sources. To illustrate this, a 10-year EPA study ending in 2001 found 45 percent of groundwaters contaminated with one or more organic chemicals. It further surveyed 26,000 industrial waste lagoons and found that one-third had no liners to prevent chemicals from leaching away. The EPA has for these reasons made cleanup of groundwater contamination a priority objective in the Superfund program.

The EPA provides polluters with resources for tackling the groundwater problem. One such resource is the Superfund Innovative Technology Evaluation Program, which currently identifies over 100 cleanup technologies and several methods for monitoring groundwaters beneath and near hazardous waste sites.

Groundwater cleanup technologies consist of physical, chemical, thermal, and bioremediation varieties, although polluted aquifers covering large areas are difficult and expensive to clean with any technology. In the 1990s the EPA's Groundwater Remediation Technologies Analysis Center published information on the most effective innovations in groundwater cleanup. By 2002 several in situ methods had come to the forefront as the best methods for removing contamination, as discussed in the previous chapter.

Few other countries have taken on the expense of groundwater cleanup even though groundwater is a major source of their drinking water. The World Health Organization (WHO) highlighted a disturbing fact in its

2006 report "Protecting Groundwater for Health." The report stated that people in poorer areas are receiving higher contaminant levels in water than those living in affluent places. Many shantytowns sit in the shadow of factories disgorging large amounts of toxic compounds into the air and water. Poor countries admit that the cost of new technologies for groundwater cleanup may be outside their reach. Groundwater pollution remains a serious and overlooked global health threat.

COSTS

Even in wealthy places like the United States, cleaning up a Superfund site requires enormous financial planning. The lengthy time necessary to clean up contamination also affects costs. In Superfund's early days, people probably assumed each NPL site cleanup would take two to three years. The jobs soon stretched to 10 to 20 years. If groundwaters were contaminated, and they usually were, the original cleanup timelines often doubled or tripled, and the costs spiraled upward.

The research group Resources for the Future forecasts that Superfund costs have reached $14 billion to $16 billion. In 1994 the Congressional Budget Committee examined the costs of Superfund's first dozen years and predicted the following: Using today's cleanup technology, inflation-corrected cleanup costs would average almost $3 billion per year through 2070. In 2004 the *New York Times* quoted EPA administrator Thomas P. Dunne as saying about Superfund that "pressures on the program's annual cleanup budget of $450 million and the growing list of sites to be restored were strangling the program's ability to operate effectively."

These forecasts demonstrate the importance of doing speedy yet thorough cleanups on contaminated sites. The ultimate goal is to delist each site from the NPL, as discussed in the sidebar "Superfund Site Delisting" on page 174.

ENVIRONMENTAL LAW

American laws can be complicated. Sometimes Congress's original intent for a new act gets lost amid conflicting views of the people it will affect: private citizens, industrialists, environmentalists. Environmental laws differ from many others in the United States because they protect not only the interests of citizens, but the Earth, plants, and animals.

SUPERFUND SITE DELISTING

Delisting is the process of removing a site from the NPL because its contamination has been cleaned up. After a polluter has rid the site of contaminants using the technologies described in this book, it arranges for another analysis of the land, air, and water. These laboratory results tell the EPA that hazardous chemicals have been reduced to acceptable levels. The polluter then sends a letter to the EPA requesting the site be removed from the NPL. The delisting process may require extra tests and one or more inspections by the EPA to the cleaned up site. In general the delisting process includes the following steps:

1. Site cleanup with the best available technologies.
2. Land, air, and water testing during and after the cleanup.
3. Submission to the EPA of testing results and a letter requesting removal from the NPL.
4. Review of project results by the local community leaders, scientists, and the EPA.
5. Site inspections and possible additional testing to confirm the cleanup results.
6. EPA decision published in the *Federal Register* and open to public comment.

A law typically begins as a bill in the House of Representatives, where it is reviewed by a committee. (Most bills do not survive the committee.) If the committee sends the bill to the rest of the House, it is debated and then voted on. If passed, it is sent to the Senate, where it goes through a similar process before the Senate sends it to the executive branch, represented by the president of the United States. The president either approves it into law or casts a veto. Some vetoed bills die, but others are overruled by Congress and become laws.

An act is a legal plan written and carried out by the government to turn a bill into a law. A law is a description of the actions (called rules of conduct) that citizens must follow to fulfill the intent of the act. People or nations follow an act but they comply with its laws. For instance, the Clean Air Act describes the requirements to be met when releasing emissions into the air. Many states have passed emissions laws to explain how to meet the air quality requirements of the Clean Air Act.

7. EPA review of public comment.
8. Delisting decision.

A designation of "Construction Completion" signifies the end of the cleanup and the elimination of health risks to humans and environment. The EPA then moves the Superfund site one column over on the NPL, from "Construction Completion" to "Site Deletion."

Plenty of Superfund sites go through this process with difficulty at almost every step. Members from the community may doubt that the site is totally free of hazards and they state their opinions during the public comment period. The EPA often responds to their concerns by deciding that the delisted site must receive extra monitoring for any further signs of pollution.

The delisting process likely works best if it remains rigorous so that polluters cannot take shortcuts in the cleanup steps. Communities should never be forced to return to the days when polluters sneaked through legal loopholes, avoiding responsibility for their actions. If the site meets all the EPA's requirements, the agency issues a Notice of Deletion. At that time the Superfund cleanup is officially complete.

Acts and laws contain legal language quite difficult for most people to understand. A regulation explains in layman terminology the detailed requirements of a law. The government agency responsible for enforcing the law also holds responsibility for writing the regulations. For example, the EPA's Office of Air and Radiation writes regulations for the Clean Air Act.

If citizens remain confused over how to comply with regulations, they may pose questions to the EPA. The EPA reviews each question and may issue a rule, which provides further explanation of the law. Finally, all related laws and regulations are numbered and organized into a single published collection called a code.

The RCRA of 1976 and the CERCLA of 1980 are the two major laws governing how waste is handled in this country. RCRA has the following three main goals:

- to protect human health and the environment from potential hazards from waste
- to ensure sound waste management practices and waste reduction
- to conserve energy and natural resources

The RCRA is intended to prevent environmental damage from pollution; CERCLA is meant to remediate pollution. CERCLA's trust fund for financing hazardous site cleanup extends not only to polluted sites in the United States, but also to those in Puerto Rico, the U.S. Virgin Islands, Guam, American Samoa, the northern Mariana Islands, and the American Trust territories in Micronesia. In summary, Superfund and the RCRA work in complementary fashion. Superfund targets hazardous waste cleanup and the RCRA focuses on preventive measures for managing hazardous and nonhazardous wastes.

Superfund's reach and budget make it the most ambitious environmental program of the past century. Other important environmental laws are listed in appendix E. Though environmental concerns continue to increase around the globe, few major environmental laws have been enacted in the United States from 1990 to 2008.

Environmental Law Terminology

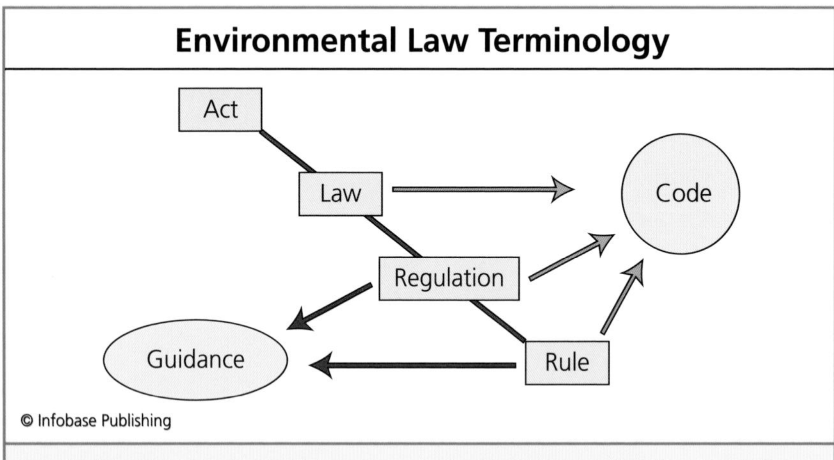

An act signed into law by the U.S. Congress or the president provides framework for the laws, regulations, and rules to be followed by U.S. residents. A guidance is not a law but rather a recommendation written by a government agency to explain how to comply with a law.

NEW CLEANUP TECHNOLOGIES

Superfund emphasizes two main points: (1) protecting human health and the environment and (2) cost-effectiveness. On occasion cleanups still must be done by the fastest method, even if it does not employ the best technology, because of local demands or immediate health threats caused by the hazards. Overall however, new technologies have been gaining ground. Different technologies have different situations in which

Innovative Technologies Used in Current Superfund Cleanups

Type of Technology	Method
For Soils and Sediments:	
biological	bioremediation, bioventing, biopiles, phytoremediation, mycofiltration
physical	air sparging/vapor extraction, soil washing, in situ flushing, grouting (solidification), dredging
chemical	oxidation, dehalogenation, solvent extraction, chemical stabilization
thermal	thermal desorption with blankets and wells
For Groundwaters:	
biological	intrinsic bioremediation, phytoremediation, air sparging/bioremediation, bioslurping
physical	pumping, grouting and walling, vacuum extraction, horizontal wells
chemical	oxidation/injection wells, electrochemical treatment, UV/oxidation, passive treatment wells
thermal	steam or hot water mobilization

they work best. Sensitive ecosystems lend themselves to natural remediation methods. Sites close to schools and homes are sometimes best fixed by in situ cleanup/treatment methods. Large sites usually manage waste removal, treatment, land restoration, and redevelopment simultaneously, as done at Fort Ord. In those cases crews clean certain parts of the land by the fast dig-and-dump method, clean other parts using innovative treatment technologies, and use natural ecosystems to restore the remaining portions of the land. Current methods for Superfund and brownfield cleanups are shown in the table on page 177.

Environmentally gentle technologies have overtaken traditional dig-and-dump methods on sites that are the original source of the hazardous

HAZARDOUS SITE CLEANUP/ TREATMENT TECHNOLOGIES, 2010

IN SITU TECHNOLOGIES	EX SITU TECHNOLOGIES
• bio- and phytoremediation	• bioremediation
• bioventing, flushing, and extraction	• extractions and soil washing
• electrical and physical separation	• thermal methods
• solidifcation/stabilization	• reduction/oxidation
• steam injection	• solidification/stabilization
• air sparging and extraction	• absorption/adsorption
• bioslurping (vacuum-enhanced bioventing)	• air stripping and water pumping
• hydrofracturing (use of pressurized water)	• oxidation and precipitation

waste. This is called *source control cleanup.* (Non-source hazardous sites are those that receive hazardous materials from another location. An oil tanker spill, for example, usually produces a non-source hazardous site.) The table on page 178 shows a comparison of ex situ and in situ methods of cleaning in 2010.

The other ex situ technologies not described by name in the table are the following: soil vapor extraction, neutralization, soil washing, soil aeration (bioventing), solvent extraction, open burning or detonation, phytoremediation, and *vitrification.* The other in situ technologies consist of thermal treatment, extraction, detoxification, phytoremediation, vitrification, and electrical separations.

The EPA's Innovative Technology Program (ITP) publishes information on new cleanup/treatment technologies now in use and also provides resources for cleanup managers seeking the most effective and least expensive solutions to a pollution problem. The main activities covered by ITP are new techniques in sampling, analysis, containment, and treatment.

CONCLUSION

The Comprehensive Environmental Response, Compensation, and Liability Act (Superfund) was enacted in 1980 in response to years of intentional and accidental waste dumping. By the 1970s several environmental catastrophes had prompted Congress to hold polluters responsible for the damage done by their discharges to the environment and human health. Today Superfund focuses on hazardous waste cleanup and is complemented by the Resource Conservation and Recovery Act, which concentrates on waste reduction and prevention. Superfund sites are those that have been assessed and found to cause an immediate serious danger to a population's health. Each of these sites is listed on a National Priorities List (NPL), which is open for public viewing and which describes the location and other details of Superfund sites.

The EPA encourages innovations for speeding assessment, planning, and cleanup and holding down costs. The Toxics Release Inventory is a database that helps speed the process by listing all environmental releases of toxic chemicals sorted by industry, company, and location. Today, several organizations provide resources for people to learn about Superfund sites located near their homes or work. The EPA provides additional information for helping cleanup managers select the appropriate technologies.

NPL sites across the country are in various stages of completion. Sometimes high costs or the sheer magnitude of the cleanup task prolongs the entire cleanup process. Some sites, like the Berkeley Pit in Montana, may never be restored to a safe condition. Groundwater contamination adds to cleanup costs because it is difficult to remove and adds complexity to a cleanup project. The EPA has strict criteria for deciding if a Superfund site has been cleaned sufficiently for it to be removed from the NPL. The final delisting process considers the needs of the community and its local government to assure a health hazard will never return to the site.

Superfund is complicated and has critics, but it nevertheless serves as the largest and most ambitious program for remediating thousands of U.S. sites contaminated by hazardous wastes.

Future Needs

The wastes in need of removal from the environment are from two sources. They are substances that accumulated long ago and they are also substances put into the air, soil, and water at present. Pollution is any solid, liquid, or gas that harms living things in the environment. Pollution is of chemical, biological, or physical form. Contamination refers to chemical or biological matter that pollutes the environment. Sometimes the magnitude of contamination at any given place is not known. Therefore contamination cleanup begins with a scientific appraisal of the hazardous materials that may be located on any site. This is called contamination assessment. It consists of studying the past history of a site, interviewing the people who lived or worked there, and doing analyses on the air, water, and soil to measure individual contaminants.

Some of the greatest strides in environmental science have taken place since the 1980s in the field of improved analytical equipment. Sensitive instruments now measure contaminants at levels as low as parts per trillion. This refined sensitivity is important for two reasons. First, it allows investigators to find each and every contaminant at a specific site. Second, environmental medicine has begun to show that some hazardous chemicals are dangerous to health in very low amounts. Being able to detect potential contaminants at very low levels therefore helps assess any health threats coming from a hazardous waste site.

After a site containing hazardous wastes has been fully assessed, cleanup begins. Today's hazardous site cleanups are conducted under the watchful eye of the EPA to assure all environmental laws are followed. Hazardous waste sites in the United States are designated as Superfund sites or as brownfields. Superfund sites are the very worst examples of air, water, and soil contamination that threaten the health of nearby humans, animals, and ecosystems. Polluters who created the hazardous conditions at a Superfund site are legally responsible for its cleanup. Brownfield sites are usually less contaminated and have greater potential than Superfund sites for being restored. Brownfield restoration is a voluntary enterprise, usually conducted by a person who wishes to develop the land for commercial use. The EPA does not actively participate in either Superfund or brownfield cleanup but it offers guidance and makes funding available to help with the cleanup. Though there have been success stories in hazardous-site cleanup, the United States still contains thousands of sites in almost every state, sites that present a potential threat to the environment.

The amount of contamination remaining to be cleaned up is formidable. Fortunately, several cleanup technologies are available. These methods draw upon some of the oldest and most basic approaches to cleanup, but they also include newer technologies with unique advantages. Excavation of hazardous substances from a dump site used to be the cheapest and easiest way to get rid of waste. It is used when a cleanup must be done fast but it can result in large piles of contaminated excavated soils. The EPA therefore encourages cleanup managers to explore more advanced technologies to reduce the final volume of waste. Many of the new technologies combine cleanup with treatment. Even excavation sites now include on-site treatment processes. Soil washing is an example of a way in which the amount of contaminated excavated soils can be minimized. In summary, almost all cleanup and treatment technologies try to reduce the amount of material to be carried away for final disposal.

When cleanup is accomplished at the waste site, it is in situ cleanup. Oxidation technology is an in situ method for cleaning up soil, water, or air. Oxidation brings the advantage of chemical reactions that are targeted at the very contaminants that must be cleaned up. It is also one of the many examples of combined cleanup/treatment. Biological cleanup methods, on the other hand, are much less specific and slower than chemical methods. But biological cleanup is desirable because it is an in situ method favored by the EPA, most communities are in favor of it, and biological processes can restore habitats once ruined by pollution. Biological cleanup can be carried out in three ways. First, microorganisms are used in a cleanup technique called bioremediation. Microorganisms may either be added to a contaminated site, or nutrients alone are added to the site to help the native microorganisms grow. By either approach bacteria and fungi degrade hazardous chemicals until they have been made harmless or are completely degraded to water and gas. Plants and trees work in a different way; their root systems absorb or adsorb contaminants. In this process, called phytoremediation, contaminants are usually extracted from the soil or stabilized in the soil, respectively. Bioremediation can be summarized as follows: (1) it works more slowly than mechanical and chemical cleanups; (2) it is less disruptive to the environment than other types of cleanup such as excavation; and (3) in general, bacteria work best on organic compounds and plants work best on metals.

The information covered in this book highlights the many choices available to cleanup managers. Excavation with heavy equipment is no longer the sole option, and chemical and biological methods bring their own advantages. But a Superfund cleanup is rarely straightforward. Almost every Superfund cleanup has been slowed by arguments over costs, funding, and legal issues. The Superfund budget has become overextended partially because a quarter of it goes to legal costs each year.

Waste cleanup has become more than choosing the best cleanup/treatment technology. More than one technology is usually deployed to meet the often conflicting objectives of industry and environmentalists. Cleanup teams must also understand environmental laws and be mindful of the economics of site cleanup. Perhaps nowhere else is economics more important than in brownfield cleanup. Brownfields are restored for the purpose of reuse. Often that reuse involves new buildings for earning income as soon as possible. Developers under pressure to begin turning a profit will choose the fastest cleanup methods available. For that reason brownfield cleanups receive input from local governments, neighbors, the community, and even the state so that as many needs as possible can be met when restoring the land.

Restoring clean water has become far more complex. Some environmentalists feel there is no body of water on Earth that has escaped some type of pollution. Tainted waters threaten the health of humans, animals, plants, and ecosystems. As the world's waters become polluted, entire countries come under severe health and economic pressure. This can lead to conflicts between countries or peoples; it already has. A large portion of the world suffers from water stress, which can cause political instability. The changing climate is expected to add to this crisis in the near future. The growing world population may soon be forced into conflicts for diminishing water supplies.

Effective technologies are available for remediating polluted waters, but water cleanup is difficult because it does not stay in one place and it moves farther than pollution on land. The best solution to preserving clean waters is to stop water pollution at its source. Point sources are easier to identify and solve than nonpoint sources. Yet point sources continue to pour contaminants into fresh surface waters, marine waters, and groundwaters. In the United States and around the world, bodies of water still receive a daily influx of intentional and accidental contamination.

Hazardous waste sites are a threat to the environment because they affect so many aspects of the environment. A single site may well contain contaminated topsoil, sediments, and surface waters. Chemicals may leach out of the sediments into an aquifer used by its neighbors for drinking water. In addition, the site may spew toxic vapors that travel in the air to affect regions far away. Environmental scientists have debunked the idea once held that hazardous wastes could be dumped or buried and then required no further thought. Almost all pollution eventually moves. Pollution does not get smaller; it spreads. Hazardous waste cleanup is therefore crucial for protecting places that have not yet been harmed by the wastes. Waste cleanup has become one of the most important aspects of today's waste management industry. Future cleanup projects will combine new technologies and standard methods. These projects will likely also depend on meaningful help from governments, legal action, and the diligence of environmentalists to assure that all dangerous substances are removed from ecosystems as quickly and thoroughly as possible.

Appendix A

PORTABLE EQUIPMENT FOR MONITORING POLLUTION			
TYPE OF ANALYZER	**HOW IT WORKS**	**WHAT IT MEASURES**	**WHERE USED**
capacitance sensor	measures waves in electric field reflected from the sample	conductivity, salinity	soil
catalytic bead sensor	measures heat emitted by gases burned on a coil coated with catalyst	combustible hydrocarbons	air
combustion gas and emissions analyzer–electrochemical sensor	electrical signal produced when gas molecules pass through a membrane	O_3, O_2, CO, CO_2, NO, NO_2, SO_2, H_2S, hydrocarbons, AsH_3, NH_3, ClO_2, HF, HCN	flue gases, air
combustion gas and emissions analyzer–infrared sensor	gas absorbs energy from infrared radiation	same as above	same as above

(continues)

PORTABLE EQUIPMENT FOR MONITORING POLLUTION *(continued)*

Type of Analyzer	How It Works	What It Measures	Where Used
combustion gas and emissions analyzer–semiconductor sensor	probe measures change in conductivity due to gas	same as above	same as above
corrosion monitor	electric probes measure metal loss by detecting resistance	metal thickness	fuel tanks, pipes, water lines, wastewater lines
density meter	sonic pulse absorbed by suspended solids	percent suspended solids, organic matter	wastewater, sludge
dissolved oxygen sensor–galvanic method	oxygen reacts with cathode to produce electrical current as it passes through membrane	dissolved oxygen	water
dissolved oxygen sensor–polarographic method	same as galvanic but uses an external electrical source	dissolved oxygen	water
fluorescence sensor	measures fluorescent light emitted from compounds when irradiated	dissolved oxygen, inorganic gases, volatile organic compounds	water

Type of Analyzer	How It Works	What It Measures	Where Used
FT-IR spectrometer (Fourier transform infrared spectroscopy)	amount of infrared light absorbed by sample is detected by sensor	organic, inorganic chemicals	soil, mine tailings
gas detectors	electrochemical sensors for specific gases	O_3, CO, Cl_2, NO, NO_2, SO_2, H_2S, AsH_3, NH_3, SiH_4, HF, HCl, HCN	air
mercury analyzer	measures atomic absorption spectrum	mercury compounds as vapors	air
photoionization detector	electrodes measure change in current due to ionized gas	volatile organic compounds	soil, water, air
suspended solids sensor	measures infrared radiation absorbed by solids	suspended solids	water
thermal conductivity sensor	measures heat transferred from a heated wire to a gas	flue gases, H_2, refrigerants (CFCs, HFCs, HCFCs)	air
trace metal analyzer–stripping voltammetry	measures energy needed to expel a metal from a coated electrode	arsenic, lead, copper, cadmium, zinc, mercury	water
trace metal analyzer–cyclic voltammetry	measures current made by metals in solution	arsenic, lead, copper, cadmium, zinc, mercury	water

(continues)

PORTABLE EQUIPMENT FOR MONITORING POLLUTION *(continued)*			
TYPE OF ANALYZER	HOW IT WORKS	WHAT IT MEASURES	WHERE USED
VOC analyzer–ionization	measures ions emitted in presence of heated diode	volatile organic compounds	air
VOC analyzer–purge and trap GC/MS (gas chromatography/mass spectrometry)	adsorbs VOCs, then releases by thermal desorption; vapors measured by GC/MS	volatile organic compounds, petroleum hydrocarbons	water
X-ray tube	measures intensity of X- ray signal from elements	chromium, titanium, barium, copper, arsenic, lead, vanadium, phosphorus, sulfur, chlorine, potassium, calcium	soil, water

Appendix B

International Organizations Involved in Pollution Control		
Organization	**Headquarters**	**Web Site**
International Atomic Energy Agency	Vienna, Austria	www.iaea.org
World Health Organization	Geneva, Switzerland	www.who.int
The World Bank	Washington, D.C., USA	www.worldbank.org
Organization for Economic Cooperation and Development	Paris, France	www.oecd.org
Worldwatch Institute (private)	Washington, D.C., USA	www.worldwatch.org
The European Union	Brussels, Belgium	europa.eu/index_en.htm
The Center for International Environmental Law	Geneva, Switzerland	www.ciel.org

(continues)

International Organizations Involved in Pollution Control *(continued)*

Organization	Headquarters	Web Site
United Nations Environment Programme	Nairobi, Kenya	www.unep.org
Commission for Environmental Cooperation	Montreal, Quebec, Canada	www.cec.org
European Environment Agency	Copenhagen, Denmark	www.eea.europa.eu

Note: All Web site addresses are active as of September 2008.

Appendix C

Microbes Used in Bioremediation		
Name of Microbe	Common Habitat	Some Hazardous Substances Degraded
Bacteria		
Achromobacter	water	chlorinated and cyclic organic compounds
Acinetobacter	water	chlorinated and cyclic organic compounds
Aeromonas	water, biofilm	various organic compounds
Arthrobacter	water	chlorinated and cyclic organic compounds
Bacillus	soil	pesticides, organic solvents, oil hydrocarbons
Beijerinckia	soil	phenanthrene
Flavobacterium	soil, water	phenanthrene
Methylosinus	soil	chlorinated and cyclic organic compounds
Moraxella	mammals	biphenyl organic compounds

(continues)

MICROBES USED IN BIOREMEDIATION *(continued)*

NAME OF MICROBE	COMMON HABITAT	SOME HAZARDOUS SUBSTANCES DEGRADED
Mycococcus	soil	chlorinated and cyclic organic compounds
Nocardia	soil	chlorinated and cyclic organic compounds
Pseudomonas	water, biofilm	chlorinated and cyclic organic compounds, bromide compounds
Xanthomonas	plants	bromide compounds
Fungi		
Candida	soil, plants	hexadecane (gasoline component)
Cladosporium	soil	decane compounds (gasoline components)
Penicillium	soil	chlorinated and cyclic organic compounds
Rhizoctonia	plants	chlorinated and cyclic organic compounds
Torulopsis	soil	oil hydrocarbons

Appendix D

Organizations that Track Environmental Hazards	
Organization/Agency	**Web Site**
Environmental Protection Agency, Envirofacts	www.epa.gov/enviro
Environmental Protection Agency, TRI	www.epa.gov/triexplorer
Green Media Toolshed	scorecard.org
Right to Know	www.rtknet.org
U.S. National Library of Medicine	toxnet.nlm.nih.gov
U.S. Geological Survey	toxics.usgs.gov/links.html
Note: All Web site addresses are active as of September 2008.	

Appendix E

Environmental Laws Enacted by the U.S. Congress	
Act, Year Signed into Law	**Administration**
Migratory Bird Treaty Act, 1918	Woodrow Wilson
Fish and Wildlife Coordination Act, 1934	Franklin D. Roosevelt
Bald Eagle Protection Act, 1940	Franklin D. Roosevelt
Federal Water Pollution Control Act, 1948	Lyndon B. Johnson
Clean Air Act, 1955	Lyndon B. Johnson
Shoreline Erosion Protection Act, 1965	Richard M. Nixon
Solid Waste Disposal Act, 1965	Richard M. Nixon
National Environmental Policy Act, 1970	Richard M. Nixon
Pollution Prevention Packaging Act, 1970	Richard M. Nixon
Resource Recovery Act, 1970	Richard M. Nixon
Coastal Zone Management Act, 1972	Richard M. Nixon
Marine Protection, Research, and Sanctuaries Act, 1972	Richard M. Nixon
Ocean Dumping Act, 1972	Richard M. Nixon

Act, Year Signed into Law	Administration
Endangered Species Act, 1973	Richard M. Nixon
Safe Water Drinking Act, 1974	Richard M. Nixon
Hazardous Materials Transportation Act, 1975	Gerald R. Ford
Resource Conservation and Recovery Act, 1976	Gerald R. Ford
Toxic Substances Control Act, 1976	Gerald R. Ford
Surface Mining Control and Reclamation Act, 1977	Jimmy Carter
Clean Water Act, 1977	Jimmy Carter
Uranium Mill-Tailings Radiation Control Act, 1978	Jimmy Carter
Comprehensive Environmental Response, Compensation, and Liability Act, 1980 (Superfund)	Jimmy Carter
Nuclear Waste Policy Act, 1982	Ronald Reagan
Emergency Planning and Community Right to Know Act, 1986	Ronald Reagan
Lead Contamination Control Act, 1988	Ronald Reagan
Medical Waste Tracking Act, 1988	Ronald Reagan
Ocean Dumping Ban Act, 1988	Ronald Reagan
Shore Protection Act, 1988	Ronald Reagan
National Environmental Education Act, 1990	George H. W. Bush
Oil Pollution Act, 1990	George H. W. Bush
Great Lakes Legacy Act, 2002	George H. W. Bush
Healthy Forests Restoration Act, 2003	George H. W. Bush
Water Resources Development Act, 2007	George H. W. Bush (vetoed)

Glossary

ACTIVATED CARBON (also *activated charcoal*) carbon granules that are especially efficient at adsorbing contaminants from air or water.

ADSORBENT a material that attracts a chemical to its outer surface and holds the chemical there.

AIR SPARGING bioremediation method by injection of air into groundwater to force chemicals out of the water as vapor.

ANTHROPOGENIC describing substances made by or only associated with human activities.

AQUIFER a natural underground source of drinking water.

BIOACCUMULATION increase in the concentration of a chemical in the tissues of organisms progressing up a food chain.

BIOAUGMENTATION addition of nutrients to the soil to encourage the growth of natural microbes that degrade toxic chemicals.

BIOCONVERSION transformation of a hazardous compound into a less hazardous chemical structure.

BIODEGRADABLE capable of being broken down by decomposers such as bacteria, fungi, and invertebrates.

BIODEGRADATION biological cleanup of hazardous wastes in the environment.

BIOFILM a mixture of microorganisms and nonliving materials that have attached to a surface in flowing liquids, such as water.

BIOGEOCHEMICAL CYCLE process that recycles Earth's nutrients in various chemical forms from the environment to living or nonliving organisms and then back to the environment.

BIOPILE an aboveground mound of contaminated soil mixed with microbes and nutrients that accelerates the degradation of toxic chemicals.

BIOREMEDIATION biological cleanup of hazardous wastes in the environment; usually refers to the activities of microorganisms.

BIOSENSOR biological device or substance that produces a reaction when it detects a specific chemical in the environment.

BIOVENTING injection of air into contaminated soil for enhancing the growth of natural microorganisms, which then degrade the contaminants.

BLEACHING a phenomenon in coral reefs in which reef organisms become harmed by unnatural water temperature, pollution, sunlight, or disease and lose pigments and die, giving the coral a white appearance.

BLOOM sudden explosive growth of algae or bacteria due to a large supply of nutrients from runoff or other sources of pollution.

BROWNFIELD abandoned or idle industrial or commercial site where redevelopment is hampered by hazardous contamination.

BT abbreviation for *Bacillus thuringiensis,* a common species of soil bacteria used in bioremediation.

BTEX benzene, toluene, ethylbenzene, and xylene; major volatile compounds in fuel hydrocarbons.

CAPPING covering of a landfill with any material that reduces odor release and prevents runoff from rain.

CATABOLISM activities in living cells or organisms to break down large molecules to smaller molecules, which can be used for energy production.

CATALYST substance that speeds a chemical or biological reaction by lowering the energy required to run the reaction.

CATALYTIC OXIDATION a reaction in which a chemical loses electrons, aided by a catalyst compound that makes the reaction energy-efficient.

CHELATOR a chemical containing a structure that enables it to form a tight, clawlike capture of a metal molecule.

COMETABOLISM a situation in which two or more microorganisms degrade a compound.

CONTAMINATION matter in air, water, or soil that causes harm or death in an organism.

CRADLE-TO-GRAVE MANAGEMENT waste management process in which the types and amounts of hazardous wastes are tracked from their release into the environment to their final disposal.

CRUDE OIL thick liquid containing mostly hydrocarbons extracted from underground oil reserves.

DDT (DICHLORODIPHENYLTRICHLOROETHANE) a chlorinated hydrocarbon pesticide, an insecticide, which has harmful long-lasting effects on the environment and is banned in the United States and some other countries.

DEHALOGENATION chemical process of removing halogen molecules (chlorine, iodine) from a compound.

DESORPTION release of a substance from a solid surface such as soil.

DETOXIFICATION rearrangement of a toxic compound's structure to reduce or eliminate its harmful effects on living tissue.

ELECTRON TRANSPORT CHAIN an energy-generating process in aerobic cells in which a series of compounds transfer electrons from one to another in a sequence that produces energy.

ENDOCRINE DISRUPTOR any synthetic chemical that when absorbed into the body interferes with hormones.

EUTROPHICATION physical, chemical, and biological changes taking place in a body of water that has received sudden high levels of nutrients, usually nitrates and phosphates.

EXCAVATION physical removal of contaminated soils from a polluted site using earthmoving equipment such as bulldozers and backhoes.

EX SITU the condition of an item being removed from its natural place.

EXTRACTION removal of a substance from soil by physical, chemical, or biological means.

EXTREMOPHILE microbe that lives in harsh conditions that other organisms cannot tolerate.

FENTON REACTIONS a complex series of linked chemical reactions involving hydrogen peroxide, a catalyst, and a chemical pollutant.

GENE SELECTION laboratory screening of many different DNA molecules to find a gene carrying a specific trait.

GENETICALLY MODIFIED ORGANISM (GMO) microorganism with DNA that has been altered by genetic engineering.

GENETIC ENGINEERING insertion of a foreign gene into an organism's DNA to give it a beneficial trait.

GLOBALIZATION a process by which commerce, communications, and travel emphasize international activities rather than domestic activities.

GREENFIELD land that has never before been developed or minimally developed with agriculture or parks.

HERBICIDE a chemical that kills weeds.

HYDROCARBON organic compound made of hydrogen and carbon atoms and a major component of fossil fuels.

HYPERACCUMULATOR plant with exceptional capacity to draw contaminants from the soil and into its roots.

INCINERATION burning process using controlled high temperatures to reduce combustible wastes to ash, water, and gas, usually carbon dioxide.

IN SITU the condition of an item being situated in its natural place.

IN SITU TREATMENT hazardous waste treatment that takes place where the contamination occurs.

INTEGRATED CLEANUP-REDEVELOPMENT the remediation of a contaminated area of land in phases in which one cleaned up section can be developed while other sections continue with pollution cleanup and removal.

INTRINSIC type of activities that are normally part of the essential nature of a thing (such as nature itself).

INTRINSIC ACTIVITY the biological actions that take place in nature without human interference, usually in reference to microbial activities.

INTRINSIC BIOREMEDIATION *also* natural attenuation; the cleanup of contaminants in soil or water due to the natural actions of microorganisms.

IN VIVO the condition of an item or organism being located in its normal place in nature.

KREBS CYCLE *also* citric acid cycle; a sequence of reactions in living cells, starting with acetic acid and ending with the production of energy stored in chemical bonds containing phosphate (PO_4).

METABOLISM activities in living cells or organisms to take in nutrients for energy for maintenance, growth, reproduction, and in some organisms, movement.

METHYLATION chemical or biological addition of a methyl group (CH_3) to a compound.

MICROBE *also* microorganism; any microscopic organism such as bacteria, protozoa, algae, or viruses.

MINERALIZATION the complete breakdown of a compound by microorganisms to water, carbon dioxide, and other simple end products.

MOMENTUM INERTIA resistance within a system to sudden change.

MTBE methyl tertiary-butyl ether, an additive in gasoline to replace lead for the purpose of reducing lead-containing car exhaust emissions.

MYCOFILTRATION process in which fungi absorb substances from soil or water.

NITRIFICATION a step in nitrogen cycling in which bacteria convert ammonium nitrogen (NH_3) to nitrate nitrogen (NO_3), making the nitrogen available for plants.

NONPOINT SOURCE an unidentified, and often multiple, point from which pollution comes, entering the environment by several routes. Examples are agricultural runoff, leaking vehicles, construction sites, and roadways.

OFF-GASES the vapors that are emitted from chemicals contained in paints, lubricants, textiles, glues, lacquers, cleaning products, and other items usually located indoors.

OXIDATION addition of an oxygen molecule to a contaminant compound to reduce its toxicity.

PATHOGEN a disease-causing microbe.

PERSISTENCE period of time in which a contaminant stays in the environment or in the body.

PESTICIDE a chemical that kills microbes or parasites.

PHYTODEGRADATION breakdown of compounds in the environment by plants.

PHYTOEXTRACTION removal of toxic chemicals from the soil by the action of plant or tree roots.

PHYTOREMEDIATION removal or neutralization of hazardous wastes in the environment through the activity of plants.

PHYTOSTABILIZATION means of keeping contaminants from moving through soils by arresting them on or in plant roots.

PHYTOTRANSFORMATION process in which plants convert toxic compounds to a nontoxic form by removing part of the chemical structure or adding something new to it.

PHYTOVOLATILIZATION process by which tree roots remove chemicals from the soil and then the tree degrades the chemicals, transports them to the leaves, and then releases the breakdown products as vapor.

PLUME single consolidated discharge of contaminants into the air or into water.

POINT SOURCE fixed location from which pollution comes such as a pipe, a smokestack, or a tanker accident.

POLLUTION physical, chemical, or biological change in air, soil, water, or food that harms human or ecosystem health.

POLYMERASE CHAIN REACTION (PCR) a technique in molecular biology for making a large amount of identical DNA from a single small piece of original DNA.

PRECIPITATE a particle that separates from a solution due to a chemical reaction or because of a physical aggregation with other materials.

PYROLYSIS the degradation of chemicals or other materials due to high heat.

RECALCITRANCE tendency of a compound to remain in the environment because it does not degrade.

REDUCTION removal of oxygen or the addition of electrons to a compound.

REFINED OIL heating oils, fuels, and lubricants made by distilling crude oil.

RELIABLE RUNOFF surface waters that can be counted on as a stable source of usable water from year to year.

RHIZOFILTRATION removal of contaminant particles from water flowing through soils by fine plant or fungal root filaments.

RIPARIAN describing an area of land adjacent to a stream, river, or wetland.

SICK BUILDING SYNDROME a situation in which harmful gases, particles, and odors accumulate indoors due to poor ventilation, causing illness to people and pets.

SOIL CREEP slow movement of soil, especially topsoil, downhill.

SOLVENT liquid that dissolves another substance to create a solution. Water is a solvent. Many solvents are organic compounds that are harmful to the environment.

SOURCE CONTROL CLEANUP methods for managing hazardous waste releases at their original source.

SUPERBUG microorganism genetically engineered to destroy contaminants in the environment.

SURFACTANT a detergent-like compound that disperses oil in water.

VITRIFICATION conversion of solids into glass; a heating process in which hazardous wastes are mixed with molten glass then cooled to form a stable solid.

VOLATILE describing any compound that quickly evaporates into the air.

VOLATILE ORGANIC COMPOUND (VOC) any organic (carbon-containing) compound that quickly evaporates into the air and harms the body; example groups of VOCs are acetones, alcohols, benzenes, and hexanes.

WATER CYCLE (also *hydrologic cycle*) process that collects, purifies, and distributes the Earth's water from the environment to living organisms and then back to the environment.

WATER STRESS situation in which water demand is greater than the available amount to meet it.

XENOBIOTIC chemical that is not normally found in nature.

Further Resources

Print and Internet

Alexander, Martin. *Biodegradation and Bioremediation.* 2d ed. San Diego: Academic Press, 1999. A technical resource on microbial and chemical reactions in waste breakdown.

American Water Works Association. "MTBE." Available online. URL: www.awwa.org/Resources/topicspecific.cfm?ItemNumber=3642&navItemNumber=32974. Accessed September 21, 2008. An overview presenting the current position on MTBE by the nation's main water utility organization.

ASARCO, Corporate Communications Department. "Montana Conservation Board Presents Environmental Award to Asarco." Business Wire, January 19, 1996. Available online. URL: www.thefreelibrary.com/Montana+Conservation+Board+Presents+Environmental+Award+to+Asarco.-a017817103. Accessed September 21, 2008. A press release covering the progress made by a polluter in rebuilding its reputation.

Barlow, Maude. "Water as Commodity—The Wrong Prescription." *Food First Backgrounder* 7, no. 3 (Summer 2001): 1–8. Available online. URL: www.foodfirst.org/en/node/57. Accessed September 21, 2008. An often-quoted article expressing the opinions of renowned Canadian water resources activist Maude Barlow.

Becker, Hank. "Phytoremediation Using Plants to Clean Up Soils." *Agricultural Research* 48, no. 6 (June 2000): 4–9. Available online. URL: www.ars.usda.gov/is/AR/archive/jun00/soil0600.pdf. Accessed September 21, 2008. An article providing a clear explanation of phytoremediation.

Biello, David. "Smog Can Make People Sick, Even Indoors." *Scientific American,* January 29, 2008. Available online. URL: www.sciam.com/article.cfm?id=smog-can-make-people-sick-even-indoors. Accessed September 21, 2008. This brief news update relates the current ozone levels in air to sick building syndrome.

Bond, Victor, ed. "Toxic Waste at Love Canal." In *American Decades: 1970–1979.* Vol. 8, *America Decades.* Farming Hills, Mich.: Gale, 1995. Available online.

URL: http://www.encyclopedia.com/doc/1G2-3468302887.html. Accessed September 18, 2008. This short article describes the health threats that resulted from toxic wastes from the 1940s.

Bourne, Joel K. "Loving Our Coasts to Death." *National Geographic,* July 2006, 60–87. An article explaining the threats to coastal ecosystems by growing coastal urban areas.

Browne, Malcolm W. "Researchers Enlist Bacteria to Do Battle with Oil Spill." *New York Times,* May 23, 1989. Available online. URL: query.nytimes.com/search/sitesearch?query=Researchers+Enlist+Bacteria+to+Do+Battle+with+Oil+Spill&date_select=full&srchst=cse. Accessed September 21, 2008. The *Exxon Valdez* oil spill and the cleanup methods being attempted.

Carson, Rachel. *The Sea around Us.* New York: Oxford University Press, 1951. One of Rachel Carson's classics on the origin and development of ocean ecosystems.

Cobb, Cathy, and Harold Goldwhite. *Creations of Fire: Chemistry's Lively History from Alchemy to the Atomic Age.* New York: Plenum Press, 1995. This book presents a detailed and interesting history of chemistry's major advances.

Daley, Beth. "State OK's Deal to Cut Bus Emissions." *Boston Globe,* December 18, 2006. Available online. URL: www.boston.com/news/local/articles/2006/12/18/state_oks_deal_to_cut_bus_emissions. Accessed September 21, 2008. An example of the pollution-control retrofitting that was initiated during Boston's Big Dig.

Daley, Richard M. "Revitalizing Chicago Through Parks and Public Spaces." Keynote address, Urban Parks Institute's "Great Parks, Great Cities" Conference, Chicago, IL, July 31, 2001. Available online. URL: www.pps.org/topics/whats_new/daley_speech. Accessed September 21, 2008. This presentation by Chicago Mayor Daley provides a historical perspective on the city's progress in hazardous waste site cleanup.

DePalma, Anthony. "Council Considers Testing Water for Traces of Drugs." *New York Times,* April 4, 2008. Available online. URL: http://www.nytimes.com/2008/04/04/nyregion/04water.html?scp=1&sq=council+considers+testing+water&st=nyt. Accessed September 21, 2008. The problem of drug contamination in New York City's drinking water.

DiGregorio, Barry E. "Unlocking *P. manganicum* Genome Could Be Bioremediation Key." *Microbe* 1, no. 11 (November 2006): 507–508. The technical aspects of using genomes in bioremediation techniques.

Doyle, Jim, and Susan Sward. "MTBE Leaks a Ticking Bomb." *San Francisco Chronicle,* December 14, 1998. Available online. URL: www.sfgate.com/cgi-bin/article.cgi?file=/chronicle/archive/1998/12/14/MN18353.DTL. Accessed

September 21, 2008. The problem of MTBE pollution in California's drinking water.

EarthSky Communications. "After IPCC, 26 Scientists Speak on Global Warming." Interview with Isaac Held, February 2, 2007. Available online. URL: www.earthsky.org/article/50989/20-scientists-speak. Accessed September 21, 2008. A variety of opinions on global warming from an impressive panel of experts on the subject.

Federal Emergency Management Agency. *Love Canal Relocation Task Force Status Report,* August 1, 1980. Available online. URL: http://ublib.buffalo.edu/libraries/specialcollections/lovecanal/documents/pdfs/lcrtf.pdf. Accessed September 21, 2008. This government report on Love Canal provides excellent insight on the thoughts of government scientists at the time of the disaster.

Feron, James. "Mt. Vernon Toxic Cleanup: A Neighborhood Disrupted." *New York Times,* January 18, 1987. Available online. URL: query.nytimes.com/gst/fullpage.html?res=9B0DE5D7163CF93BA25752C0A961948260&scp=1&sq=Mt.+Vernon+Toxic+Cleanup%3A+A+Neighborhood+Disrupted&st=nyt. Accessed September 21, 2008. An article describing a toxic waste cleanup in New York City in which one remediation method used oxidation.

Fox, Robert D. "Recipe for Revitalization: Key Ingredients for Brownfields Redevelopment." Manko, Gold, Katcher, & Fox, 2001. This short article covers the financial benefits of brownfield cleanup.

Gilliom, Robert J., Jack E. Barbash, Charles G. Crawford, Pixie A. Hamilton, Jeffrey D. Martin, Naomi Nakagaki, Lisa H. Nowell, et al. *Pesticides in the Nation's Streams and Ground Water, 1992–2001.* Rev. ed. Reston, Va.: U.S. Geological Survey, 2007. Available online. URL: http://pubs.usgs.gov/circ/2005/1291/. Accessed September 20, 2008. A lengthy but very readable report on pesticides and water pollution.

Gold, Allan R. "Ideas and Trends; A Company Looks for a Cheap, Natural Way to Clean Up Pollution." *New York Times,* January 27, 1991. Available online. URL: http://query.nytimes.com/gst/fullpage.html?res=9D0CE3D71F30F934A15752C0A967958260&scp=1&sq=Ideas+and+Trends%3B+A+Company+Looks+for+a+Cheap%2C+Natural+Way+to+Clean+Up+Pollution&st=nyt. Accessed September 21, 2008. This article describes biological cleanup of PCBs using bacteria.

Goldbaum, Ellen. "Micro-organisms Genetically Engineered into Tiny Factories." *Medicalnewstoday.com,* September 21, 2007. Available online. URL: http://www.medicalnewstoday.com/articles/83006.php. Accessed September 21, 2008. An article presenting insight into the potential uses of bioengineered bacteria.

Graham, Sarah. "Plants Dispatched to Decontaminate Soil." *Scientific American,* April 12, 2004. Available online. URL: www.sciam.com/article.cfm?id=plants-dispatched-to-deco. Accessed September 21, 2008. This short article gives a concise explanation of phytoremediation.

Gutierrez, Hector. "Attorney General Says Rocky Mountain Arsenal Staying Contaminated Past Cleanup Date." *Rocky Mountain News,* October 30, 2007. Available online. URL: http://www.rockymountainnews.com/drmn/local/article/0,1299,DRMN_15_5734994,00.html. Accessed September 21, 2008. A news article relaying the current cleanup timetable for Rocky Mountain Arsenal and potential future threats to the environment.

Handwerk, Brian. "Plants Perform 'Green Clean' of Toxic Sites." *Nationalgeographic.com/news,* September 24, 2004. Available online. URL: news.nationalgeographic.com/news/pf/95746439.html. Accessed September 21, 2008. An article on the promise of phytoremediation for toxic cleanup.

Hightower, Mike, and Suzanne A. Pierce. "The Energy Challenge." *Nature* 452 (March 20, 2008): 285–286. Available online. URL: http://www.nature.com/nature/journal/v452/n7185/full/452285a.html. Accessed September 28, 2008. An excellent article on the political aspects of energy generation and water use worldwide.

Holmberg, David. "At Some Superfund Sites, Toxic Legacies Linger." *New York Times,* January 20, 2008. Available online. URL: http://www.nytimes.com/2008/01/20/nyregion/nyregionspecial2/20Rsuperfund.html?scp=1&sq=At+some+Superfund+sites%2C+toxic+legacies+linger&st=nyt. Accessed September 21, 2008. The author describes a walkthrough of a New Jersey toxic site and the challenges of Superfund cleanup.

Ingold, John. "'Dew of Death' Discovery Shuts Wildlife Refuge." *Denver Post,* November 2, 2007. Available online. URL: http://denverpost.com/news/ci_7346675. Accessed September 21, 2008. This news article recounts a disturbing discovery of hidden toxic chemicals at a wildlife refuge.

International Water Association. *The Bonn Charter for Safe Drinking Water,* September 2004. Available online. URL: docs.watsan.net/Downloaded_Files/PDF/IWA-2004-Bonn.pdf. Accessed September 20, 2008. A booklet outlining the Bonn Charter in general terms.

Jackson, Jeremy. "Brave New Ocean." Roger T. Peterson Memorial Lecture, presented at Harvard University, Cambridge, Mass., April 6, 2008. This lecture was presented by a preeminent oceanographer on the current degradation of the world's ocean ecosystems.

James, Susan Donaldson. "Love Canal's Lethal Legacy Persists." *ABC News,* August 11, 2008. Available online. URL: abcnews.go.com/Health/story?id=5553393.

Accessed September 18, 2008. A network news story recounting the accident at Love Canal and its health effects thirty years later.

Janofsky, Michael. "Changes May Be Needed in Superfund, Chief Says." *New York Times,* December 5, 2004. Available online. URL: www.nytimes.com/2004/12/05/politics/05super.html?_r=1&scp=1&sq=Changes+May+Be+Needed+in+Superfund%2C+Chief+Says&st=nyt&oref=slogin. Accessed September 21, 2008. This article describes the political challenges of managing Superfund's enormous budget.

Johnson, Kirk. "Two Years Later: Air Quality; Study Says Ground Zero Soot Lingered." *New York Times,* September 11, 2003. Available online. URL: query.nytimes.com/gst/fullpage.html?res=9500EFD9123BF932A2575AC0A9659C8B63&scp=1&sq=Two+Years+Later%3A+Air+Quality%3B+Study+Says+Ground+Zero+Soot+Lingered&st=nyt. Accessed September 21, 2008. This article reexamines the air pollution problem caused by the World Trade Center attack in 2001.

Joling, Dan. "Foes Doubt Oil Industry Cleanup Plans." *International Business Times,* April 11, 2008. Available online. URL: www.ibtimes.com/articles/20080411/foes-doubt-oil-industry-cleanup-plans.htm. Accessed September 28, 2008. This article discusses the problems associated with cleaning oil spills from ice floes off Alaska's coast.

Jowit, Juliette. "Plastic Waste Threat to Marine Life." *Guardian Observer,* September 16, 2007. Available online. URL: www.guardian.co.uk/environment/2007/sep/16/pollution.travelnews. Accessed September 19, 2008. An overview of plastic contamination of oceans worldwide.

Leeth, Dan. "Colorado Autumn: Gold Digging." *Denver Post,* September 9, 2007. Available online. URL: www.denverpost.com/ci_6823105. Accessed September 21, 2008. A travel article that gives a short synopsis of Colorado's Mineral Belt Trail.

Maag, Christopher. "In the Battle against Cancer, Researchers Find Hope in a Toxic Wasteland." *New York Times,* October 9, 2007. Available online. URL: www.nytimes.com/2007/10/09/us/09pit-ml?_r=1&scp=1&sq=Researchers+Find+Hope+in+a+Toxic+Wasteland+&st=nyt&oref=slogin. Accessed September 28, 2008. An article describing the studies on microbes living in the Berkeley Pit in Montana.

Madigan, Michael T., and Barry L. Marrs. "Extremophiles." *Scientific American,* April 1997. Available online. URL: atropos.as.arizona.edu/aiz/teaching/a204/extremophile.pdf. Accessed September 28, 2008. This is an excellent article with good illustrations.

Magnuson, Ed. "The Poisoning of America." *Time,* 22 September 1980. Available online. URL: www.time.com/time/printout/0,8816,952748,00.html#.

Accessed September 21, 2008. This in-depth report exposes the pollution problems first being recognized in 1980.

Mahr, Krista. "Trying to Save the Coral Reefs." *Time,* 17 August 2007. Available online. URL: http://www.time.com/time/health/article/0,8599,1653804,00.html. Accessed September 21, 2008. A summary of the current state of the world's corals and their monitoring.

Massachusetts Turnpike Authority. "Controlling Diesel Air Pollution." *Massturnpike.com,* 2008. Available online. URL: http://www.massturnpike.com/bigdig/background/airpollution.html. Accessed September 21, 2008. A public relations article presenting the Big Dig's program for controlling air pollution from heavy equipment and trucks.

May, Meredith. "Hair and Mushrooms Create a Recipe for Cleaning Up Oily Beaches." *San Francisco Chronicle,* November 14, 2007. Available online. URL: http://www.sfgate.com/cgi-bin/article.cgi?f=/c/a/2007/11/14/MNPQTBLE4.DTL. Accessed September 21, 2008. A description of a unique biological cleanup method.

McWhinney, Jim. "Water: The Ultimate Commodity." *Investopedia.com,* 2008. Available online. URL: http://www.investopedia.com/articles/06/Water.asp. Accessed September 21, 2008. An article describing a unique method for cleaning up oil spill-marred beaches.

Merrero, Tony. "Commissioners Reluctant to Up Contribution to Water Authority." *Hernando Today: The Tampa Tribune,* March 19, 2008. Available online. URL: http://www2.hernandotoday.com/content/2008/mar/19/ha-commissioners-reluctant-to-up-contribution-to-w. Accessed September 21, 2008. An article describing a local groundwater shortage threatening a Florida county.

Miller, G. Tyler. *Environmental Science: Working with the Earth.* Belmont, Calif.: Thomson Learning, 2006. An excellent resource for environment science with a good glossary and excellent illustrations.

Miller, Jonathan. "The New Toxic-Site Cleanup Agent: A Bacterium that Gobbles Up Poison." *New York Times,* October 19, 2003. Available online. URL: http://query.nytimes.com/gst/fullpage.html?res=9D0DE3D6133EF93AA25753C1A9659C8B63&scp=1&sq=The+New+Toxic-site+Cleanup+Agent%3A+A+Bacterium+that+Gobbles+Up+Poison&st=nyt. Accessed September 21, 2008. A recounting of the discovery of a bacterial species that degrades toxic organic compounds.

National Resources Defense Council. "Rep. Bass Putting MTBE Polluters Before People." *NRDC.org,* July 18, 2005. Available online. URL: www.nrdc.org/media/pressreleases/050718a.asp. Accessed September 21, 2008. The environmental organization's news release on U.S. Congress's discussions of the MTBE pollution problem.

Nesbitt, Jamie. "Woman on the Move: Rosanne Sanchez." *Brownfield News,* October 2007. This magazine article presents a general update on brownfield restoration.

Pacific Northwest National Laboratory. "Supercritical Water Oxidation/Synthesis." July 24, 2003. Available online. URL: www.pnl.gov/supercriticalfluid/tech_oxidation.stm. Accessed September 21, 2008. A short but descriptive article on this new technology.

Palmer, Douglas. "Letter from President Palmer." *Recycling America's Land: 2008 Brownfields Survey.* U.S. Conference of Mayors. Available online. URL: http://www.usmayors.org/brownfields/brownfields_bp08.pdf. Accessed September 19, 2008. The mayor of Trenton, New Jersey, gives the opening comments for a technical report on brownfield developments to date.

Pulaski, Alex, and Julie Sullivan. "One Cleanup Shapes Another," *Oregonian,* May 19, 2007. A news article describing aspects of Oregon's Willamette River toxic cleanup.

Quillen, Ed. "The Last Mine Closes in Leadville." *High Country News,* February 15, 1999. Available online. URL: www.hcn.org/servlets/hcn.Article?article_id=4782. Accessed September 21, 2008. An overview of Leadville, Colorado's mining industry.

Qurik, James A. "Frogs Have Returned, But Much Work Remains." *Asbury Park Press,* November 22, 2004. Available online by subscription. URL: http://nl.newsbank.com/nl-search/we/Archives?s_site=app&f_site=app&f_sitename=Asbury+Park+Press+%28Neptune%2C+NJ%29&p_theme=gannett&p_product=ASBB&p_action=search&p_field_base-0=&p_text_base-0=Frogs+have+returned+but+much+work+remains&Search=Search&p_perpa ge=10&p_maxdocs=200&p_queryname=700&s_search_type=keyword&p_sort=_rank_%3AD&p_field_date-0=YMD_date&p_params_date-0=date%3AB%2CE&p_text_date-0=. The story of a flawed toxic cleanup in New Jersey.

Rai, U.N., and Amit Pal. "Toxic Metals and Phytoremediation." *EnviroNews: Newsletter of ISEB India* 5, no. 4 (October 1999). Available online. URL: isebindia.com/95_99/99-10-2.html. Accessed September 21, 2008. A clear overview of types of phytoremediation.

Rain—Public Internet Broadcasting. "The Oil Spill of 1969." *Rain.org,* 2004. Available online. URL: http://www.rain.org/campinternet/channelhistory/oil/oilspill.html. Accessed September 21, 2008. This brief article describes an oil spill that fouled the coast of Santa Barbara, California.

Riverkeeper, Inc. "Hudson Fisheries: the Facts. Hudson River Power Plant Fish Kills." 2008. Available online. URL: www.riverkeeper.org/campaign.php/biodiversity/the_facts/178. Accessed September 21, 2008. An environmental group describes fish kills in New York's Hudson River due to thermal pollution.

Schmoll, Oliver, Guy Howard, John Chilton, and Ingrid Chorus. *Protecting Groundwater for Health: Managing the Quality of Drinking-Water Sources.* London: IWA Publishing, 2006. Information on how groundwaters are contaminated by chemicals and microbes, and includes clean-water strategies.

Schneider, Keith. "Ideas and Trends; Man to Microbe: Do a Job, Drop Dead." *New York Times,* March 27, 1988. Available online. URL: query.nytimes.com/gst/fullpage.html?res=940DE1DE103EF934A15750C0A96E948260&scp=1&sq=Ideas+and+Trends%3B+Man+to+Microbe%3A+Do+a+Job%2C+Dro p+Dead&st=nyt. Accessed September 21, 2008. A science article describing how suicide genes work in bioengineering.

Science Daily. "Gene Transfer between Species Is Surprisingly Common." March 11, 2007. Available online. URL: www.sciencedaily.com/releases/2007/03/070308220454.htm. Accessed September 21, 2008. Describes recent findings on gene transfer between bacteria.

Seech, Allan, and James Mueller. "Pesticide Terminator." *Environmental Protection,* May 1, 2007. Available online. URL: www.eponline.com/articles/54418. Accessed September 21, 2008. A review of bioremediation for pesticide-contaminated soils.

Seipel, Tracy. "Mine Companies to Pay State, EPA $8 Million in Costs Incurred at Leadville." *Denver Post,* June 26, 1993. Available online by subscription. URL: nl.newsbank.com/nl-search/we/Archives. Accessed September 21, 2008. This article covers the payment by the creator of the Leadville, Colorado, toxic waste.

Severo, Richard. "Three Mile Island Cleanup Comes to a Virtual Standstill as Experts Ponder; Middleton, PA." *New York Times,* January 6, 1981. Available online. URL: query.nytimes.com/gst/fullpage.html?sec=health&res=9407E7DD173BF935A35752C0A967948260&scp=1&sq=Three+Mile+Island+Cleanup+Comes+to+a+Virtual+Standstill+as+Experts+Ponder%3B+Middleton%2C+PA&st=nyt. Accessed September 21, 2008. A news article that updates the work and planning under way for cleaning up a radioactive material accident.

Sheahan, Judy, and Derrick Coley. "Historic Brownfields Bill Becomes Law." *U.S. Mayor Newspaper,* January 14, 2002. Available online. URL: http://www.usmayors.org/uscm/us_mayor_newspaper/documents/01_14_02/

brownfields1.asp/. Accessed September 21, 2008. An excellent review of the U.S. brownfields program.

Stone, Gregory S. "Deep Science: Sleeping with the Fishes." *Scientific American,* September 2003. The author describes his stay aboard a deep-sea research vessel.

Sullivan, Michael. "Australia Turns to Desalination Amid Water Shortage." *Morning Edition, National Public Radio,* June 18, 2007. Available online. URL: http://www.npr.org/templates/story/story.php?storyId=11134967. Accessed September 21, 2008. A news story describing desalination operations in Perth, Australia.

Sullivan, Tim. "Current of Anger in the Ganges." *Los Angeles Times,* March 18, 2007. This article covers the water pollution crisis in India and the challenges to cleanup.

Taghari, Safiyh, Tanja Barac, Bill Greenberg, Brigitte Borremans, Jaco Vangronsveld, and Daniel van der Lelie. "Horizontal Gene Transfer to Endogenous Endophytic Bacteria from Poplar Improves Phytoremediation of Toluene." *Applied and Environmental Microbiology* 71, no. 12 (December 2005): 8500–8505. A technical journal article on bioengineering of phytoremediation plants.

Templeton, David. "Cleaner Air Is Legacy Left by Donora's Killer 1948 Smog." *Pittsburgh Post-Gazette,* October 29, 1998. Available online. URL: www.post-gazette.com/magazine/19981029smog1.asp. Accessed September 21, 2008. An excellent recounting of an air pollution disaster that occurred in 1948.

U.S. Department of State. "The Superfund Helps Clean Up Hazardous Waste." *America.gov,* March 12, 2008. Available online. URL: www.america.gov/st/env-english/2008/March/20080312165549wrybakcuh0.3213421.html. Accessed September 21, 2008. A brief but clear synopsis of Superfund.

U.S. Environmental Protection Agency. "Leadville, Colorado: Moving Beyond the Scars of Mining, Integrating Remedial Design and Site Reuse." Available online. URL: www.epa.gov/superfund/programs/recycle/pdf/cal_gulch.pdf. Accessed September 21, 2008. A detailed and readable brochure on the history of Leadville, Colorado's mining and the development of the Mineral Belt Trail.

U.S. Environmental Protection Agency. "An Environmental Revolution." *History, EPA.gov,* September 21, 2007. Available online. URL: www.epa.gov/history/publications/origins5.htm. Accessed September 21, 2008. A brief history of the National Environmental Policy Act.

U.S. Environmental Protection Agency. Office of Emergency and Remedial Response, Oil Program Center. "Understanding Oil Spills and Oil Spill

Response." 1999. Available online. URL: www.epa.gov/OEM/content/learning/pdfbook.htm. Accessed September 20, 2008. An excellent primer on the effects of oil spills on the environment, ecosystems, and wildlife.

U.S. Environmental Protection Agency. Office of Radiation and Indoor Air. "The Inside Story: A Guide to Indoor Air Quality." 1993; updated 2008. Available online. URL: www.epa.gov/iaq/pubs/insidest.html. Accessed September 21, 2008. This government publication provides valuable information on all aspects of indoor air pollution.

U.S. Environmental Protection Agency. Office of Water. "National Water Quality Inventory: Report to Congress, 2002 Reporting Cycle." October, 2007. Available online URL: www.epa.gov/305b/2002report. Accessed September 28, 2008. A lengthy and detailed resource on freshwater quality.

U.S. Environmental Protection Agency. Technology Innovation Office. "A Citizen's Guide to Bioremediation." April 2001. Available online. URL: www.epa.gov/tio/download/citizens/bioremediation.pdf. Accessed September 21, 2008. This brochure draws a simplified overview on bioremediation.

U.S. House of Representatives. 1979. House Sub-Committee on Oversight and Investigations. *Testimony presented to the House Sub-Committee on Oversight and Investigations: Lois M. Gibbs, President, Love Canal Homeowners Association.* 96th Cong., 1st sess., March 21. Available online. URL: ublib.buffalo.edu/libraries/specialcollections/lovecanal/documents/pdfs/gibbs.pdf. Accessed September 21, 2008. The riveting testimony of Lois Gibbs, who initiated government action in the Love Canal crisis.

Williams, Ted. "Drunk on Ethanol." *Audubon,* August 2004. Available online. URL: magazine.audubon.org/incite/incite0408.html. Accessed September 21, 2008. This opinion piece covers the pitfalls of corn production for ethanol biofuel production.

Woolley, John, and Gerhard Peters. "Richard Nixon. 353–Veto of the Federal Water Pollution Control Act Amendments of 1972, October 17." American Presidency Project, University of California at Santa Barbara. Available online. URL: www.presidency.ucsb.edu/ws/?pid=3634. Accessed September 21, 2008. The verbatim veto from President Nixon of a bill from the Congress on financing fresh surface water cleanup.

World Health Organization/UNICEF. "Water for Life: Making It Happen." Geneva: World Health Organization, 2005. Available online. URL: www.

who.int/water_sanitation_health/monitoring/jmp2005/en/index.html. Accessed on September 21, 2008. One of several annual reports on worldwide water supply, crises, and quality as related to health and disease.

World Meteorological Organization. "WMO Marks World Meteorological Day with Call to Strengthen Climate Observations." Press release no. 811, March 23, 2008. Available online. URL: www.wmo.ch/pages/mediacentre/press_releases/pr_811_en.html. Accessed September 21, 2008. An interesting update on climate-related occurrences worldwide.

Web Sites

Greenpeace USA. Available online. URL: www.greenpeace.org/usa. Accessed September 21, 2008. Provides statistics and current issues on the environment.

The Groundwater Foundation. Available online. URL: www.groundwater.org. Accessed September 21, 2008. An excellent educational site for students on the subject of groundwater.

International Maritime Organization. Available online. URL: www.imo.org. Accessed September 21, 2008. International organization focused on protecting the health of the oceans for commerce and for environment.

Organization for Economic Cooperation and Development. Available online. URL: www.oecd.org. Accessed September 21, 2008. Comprehensive international programs on sustainable living, commerce, and environmental protection.

Portland Cement Association. Available online. URL: www.cement.org. Accessed September 21, 2008. Industry association provides resources on stabilization and solidification, sustainable building, and Superfund statistics.

United Nations Environment Programme. Available online. URL: www.unep.org. Accessed September 21, 2008. Covers international laws on environmental pollution, water conservation, pollution, and waste management.

U.S. Environmental Protection Agency. Available online. URL: www.epa.gov/. Accessed September 21, 2008. Covers every aspect of pollution cleanup technology, environmental law, and new technologies for reducing waste, plus databases on hazardous chemicals and hazardous sites.

U.S. Geological Survey. Available online. URL: www.usgs.gov. Accessed September 21, 2008. A resource for the current condition of the land and waters. Includes publications on environmental topics.

World Health Organization. Available online. URL: www.who.int/en. Accessed September 21, 2008. Covers global pollution in relation to human disease and mortality.

World Meteorological Organization. Available online. URL: www.wmo.ch/pages/index_en.html. Accessed September 21, 2008. International resource on climate change and water pollution.

World Water Council. Available online. URL: www.worldwatercouncil.org. Accessed September 21, 2008. Resource on water quality's effect on international relations and health.

Index

Note: Page numbers in *italic* refer to illustrations. The letter *t* indicates tables.

C

D

I

J

K

L

M